ESSENTIAL

MATHEMATICS STAGE 7

FOR CAMBRIDGE SECONDARY 1

Patrick Kivlin, Sue Pemberton, Paul Winters

Oxford excellence for Cambridge Secondary 1

OXFORD
UNIVERSITY PRESS

Great Clarendon Street, Oxford, OX2 6DP, United Kingdom

Oxford University Press is a department of the University of Oxford.
It furthers the University's objective of excellence in research, scholarship,
and education by publishing worldwide. Oxford is a registered trade mark of
Oxford University Press in the UK and in certain other countries

British Library Cataloguing in Publication Data
Data available

978-1-4085-1983-7

10 9 8

Printed in Great Britain by CPI Group (UK) Ltd., Croydon CR0 4YY

Acknowledgements

Cover: aricvyhmeister/iStockphoto
Illustrations: OKS Prepress, India and Angela Knowles
Page make-up: OKS Prepress, India

The publishers would like to thank the following for permission to use their photographs:

P1: ZU_09/iStockphoto; **p20:** leminuit/iStockphoto; **p32:** benkrut/iStockphoto; **p46 l:** adamkaz/iStockphoto;
p46 mid l: stanley45/iStockphoto; **p46 mid r:** shuchunke/iStockphoto; **p46 r:** Alextom2k/iStockphoto; **p50f:**
alexjuve/Fotolia; **p50g:** Santi/Fotolia; **p54:** samgrandy/iStockphoto; **p68:** bruce7/iStockphoto; **p82:** Keith
Foster/Alamy **p84 l:** JohnnyGreig/iStockphoto; **p84 r:** skynesher/iStockphoto; **p99 top r:** uchar/iStockphoto;
p99 mid l: Devonyu/iStockphoto; **p99 mid centre:** malerapaso/iStockphoto; **p99 mid r:** flyfloor/iStockphoto;
p99 bottom l: rambo182/iStockphoto; **p99 bottom r:** Henrik5000/iStockphoto; **p100 top:** jondpatton/
iStockphoto; **p100 bottom:** mgkaya/iStockphoto; **p101 top:** Mosutatsu/iStockphoto; **p101 mid l:** kertlis/
iStockphoto; **p101 mid r:** drbimages/iStockphoto; **p101 bottom:** dlerick/iStockphoto; **p102 top** tunart/
iStockphoto; **p102 bottom l:** Agorohov/iStockphoto; **p102 bottom cen:** xefstock/iStockphoto; **p102 bottom
r:** posteriori/iStockphoto; **p103 top l:** mgkaya/iStockphoto; **p103 top mid:** AnthiaCumming/iStockphoto;
p103 top r: Turnervisual/iStockphoto; **p103 mid l:** aziatik13/Fotolia; **p103 mid r:** Halfpoint/Fotolia; **p103
bottom:** Henrik5000/iStockphoto; **p105 top l:** Difydave/iStockphoto; **p105 top r:** AnthonyRosenberg/
iStockphoto; **p105 bottom l:** ia_64/iStockphoto; **p105 bottom mid l:** JuSun/iStockphoto; **p105 bottom
mid r:** kbwills/iStockphoto; **p105 bottom r:** SteveMcsweeny/iStockphoto; **p111:** EtiAmmos/Shutterstock;
p126, 127, 131, 133, 137: williammpark/iStockphoto; **p157:** Andrew_Howe/iStockphoto; **p162:** davidhills/
iStockphoto; **p169:** travellinglight/iStockphoto; **p175:** adventtr/iStockphoto; **p182:** davidevison/iStockphoto;
p200: DGLowrie/iStockphoto; **p208:** RapidEye/iStockphoto; **p221:** Tolga_TEZCAN/iStockphoto; **p232:**
brankokosteski/iStockphoto; **p243:** Matt_Brown/iStockphoto; **p255:** PaulCowan/iStockphoto; **p266:** evemilla/
iStockphoto; **p267:** PhotoTalk/iStockphoto; **p272:** 4x6/iStockphoto; **p277:** kikkerdirk/iStockphoto; **p286 top:**
derrrek/iStockphoto; **p286 bottom:** R-J-Seymour/iStockphoto

Although we have made every effort to trace and contact all
copyright holders before publication this has not been possible in all
cases. If notified, the publisher will rectify any errors or omissions at
the earliest opportunity.

Contents

	Introduction	iv
1	Integers, powers and roots	1
2	Expressions	20
3	Shapes and geometric reasoning 1	32
4	Fractions	54
5	Decimals	68
6	Processing, interpreting and discussing data	84
7	Length, mass and capacity	99
8	Equations	111
9	Shapes and geometric reasoning 2	121
10	Presenting, interpreting and discussing data	143
11	Area, perimeter and volume	157
12	Formulae	175
13	Position and movement	182
14	Sequences	200
15	Probability	208
16	Functions and graphs	221
17	Fractions, decimals and percentages	232
18	Planning and collecting data	243
19	Ratio and proportion	255
20	Time and rates of change	266
21	Sets (extension work)	277
22	Matrices (extension work)	286
	Glossary	293
	Index	297

Introduction

Welcome to *Mathematics for Cambridge Secondary 1!* This student book has been written for the Cambridge International Examinations Secondary 1 Mathematics Curriculum Framework and provides complete coverage of Stage 7. Created specifically for international students and teachers by a dedicated and experienced author team, this book covers all areas in the curriculum: number, algebra, geometry measure, handling data and problem solving.

The following features have been designed to guide you through the content of the book with ease:

Learning outcomes

The learning outcomes give you an idea of what you will be covering, and what you should understand by the end of the chapter.

Worked examples

Worked examples to illustrate and expand the content. Work through these yourself and then compare your answers with the solutions.

Problem solving questions: Help develop knowledge and skills by requiring creative or methodical approaches, often in a real-life context.

Extension questions: Provide you with further challenge beyond the standard questions found in the book.

Hints: Useful tips for you to remember whilst learning the maths.

Key words: The first time key words appear in the book, they are highlighted in bold red text. A definition of each key word can be found in the glossary at the back of the book.

Extension Chapters: Chapters 21 and 22 are included in the Cambridge IGCSE curriculum. This additional content is included to challenge more able students. (Note that the route matrix work is not examined.)

This book is part of a series of six books and three teacher-support CD-ROMs. There are three student textbooks covering stages 7, 8 and 9 and three workbooks written to closely match the textbooks, as well as a teacher's CD-ROM for each stage.

The accompanying **Workbooks** provide extensive opportunities for you to practice your skills and apply your knowledge, both for homework and in the classroom.

The **teacher's CD-ROMs** include a wealth of interactive activities and supplementary worksheets, as well as the answers to questions in the books.

1 Integers, powers and roots

Learning outcomes

- Recognise multiples, factors, common factors and primes.
- Find the lowest common multiple in simple cases.
- Make use of simple tests of divisibility.
- Use the 'sieve' for generating primes developed by Eratosthenes.
- Recognise squares of whole numbers to at least 20×20 and the corresponding square roots.
- Use the notation 7^2 and $\sqrt{49}$.
- Recognise negative numbers as positions on a number line, and order, add and subtract positive and negative integers in context.

Think of a number

The study of different types of number goes back many years to many different civilisations. In both Greece and Egypt important results about prime numbers were being discovered more than 2000 years ago.

Euclid, a Greek mathematician working around 300 BCE, proved that the list of prime numbers goes on forever.

More recently, the hunt for bigger and bigger prime numbers has been helped by the use of computers.

Very large prime numbers are used to create secure codes, which are used in internet banking.

1.1 Multiples

Look at these multiplications.

$$1 \times 8 = \mathbf{8} \qquad 2 \times 8 = \mathbf{16} \qquad 3 \times 8 = \mathbf{24} \qquad 4 \times 8 = \mathbf{32} \qquad 5 \times 8 = \mathbf{40} \qquad 6 \times 8 = \mathbf{48}$$

The numbers 8, 16, 24, 32, 40, 48 are multiples of 8.

The multiples of 8 do not stop at 48. You could carry on with the series and find more.

$7 \times 8 = \mathbf{56}$ 56 is the 7th multiple of 8.

$50 \times 8 = \mathbf{400}$ 400 is the 50th multiple of 8.

Sometimes you will use the inverse calculation to check.

An inverse is the opposite of an operation.

If you do an operation and then the inverse, you go back to the start.

For example, the inverse of addition is subtraction.

If you add 20 and then subtract 20, there is no change to the original number.

This table shows the inverse of each type of calculation.

Original operation	Inverse operation
Addition	Subtraction
Subtraction	Addition
Multiplication	Division
Division	Multiplication

Worked example 1

Write down the inverse of each operation.

a ×15   **b** +40   **c** ÷8   **d** −32

Using the table above helps.

a The inverse of ×15 is ÷15

b The inverse of +40 is −40

c The inverse of ÷8 is ×8

d The inverse of −32 is +32

Worked example 2

a Write down the 10th multiple of 8.

b Use multiplication to find the 28th multiple of 8.

c Use division to show that 584 is a multiple of 8.

a The 1st multiple of 8 is 1 × 8; the 2nd multiple of 8 is 2 × 8, etc.

The 10th multiple of 8 is 10 × 8 = **80**

b The 28th multiple of 8 will be 28 × 8

Remember how to work this out:

$$\begin{array}{r} 28 \\ \times \ \ 8 \\ \hline 22\underset{6}{4} \end{array}$$

> Another way to check this is to find the 30th multiple as 30 × 8 = 240
>
> Then subtract 16 to find the 28th multiple. 240 − 16 = 224

The 28th multiple of 8 is 224.

c To show that 584 is a multiple of 8 you need to
   work out 584 ÷ 8

$$\begin{array}{r} 0\ 7\ 3 \\ 8\overline{)5\ {}^58\ {}^24} \end{array}$$

> Remember that division is the opposite of multiplication.

There is no remainder so 73 × 8 = 584

This shows that 584 is a multiple of 8.

Common multiples

Multiples of 6 are 6, 12, 18, **24**, 30, 36, 42, **48**, 54, …

Multiples of 8 are 8, 16, **24**, 32, 40, **48**, 56, 64, 72, …

24 and 48 are in both lists. These are common multiples of 6 and 8.

The next common multiple would be 72, then 96, etc.

24 is the smallest common multiple and it is known as the lowest common multiple of 6 and 8.

Tests for divisibility

In part **c** of Worked example 2, you had to check if a given number was divisible by 8.

For some numbers there is a way of checking divisibility.

Some numbers are easier than others to check.

Number	Test for divisibility
2	All even numbers are divisible by 2. The final digit of an even number is 0, 2, 4, 6 or 8.
3	Add up the digits in the number. The number is divisible by 3 if the sum of the digits is also divisible by 3.
4	A number is divisible by 4 if the final two digits are a multiple of 4.
5	A number is divisible by 5 if the final digit is either 5 or 0.
6	A number is divisible by 6 if it is divisible by 2 and also by 3. Use both tests to check.
7	There is no easy test for divisibility by 7.
8	A number is divisible by 8 if the final three digits are a multiple of 8.
9	Add up the digits in the number. The number is divisible by 9 if the sum of the digits is also divisible by 9.
10	A number is divisible by 10 if the final digit is 0.
100	A number is divisible by 100 if the final two digits are 00.

Worked example 3

Which of the following numbers divide exactly into 28 548?
2, 3, 4, 5, 6, 8, 9,10

Check each number in turn.

2 The last digit is 8, so it is an even number.

28 548 is divisible by 2.

3 $2 + 8 + 5 + 4 + 8 = 27$

27 is a multiple of 3. ● ─── You can repeat this if necessary, $2 + 7 = 9$ so 9 is a multiple of 3.

28 548 is divisible by 3.

4 The last two digits are 48. This is a multiple of 4.

28 548 is divisible by 4.

5 The final digit is 8, so 28 548 is not divisible by 5.

6 The final digit is 8. It divides by 2.

$2 + 8 + 5 + 4 + 8 = 27$

27 is a multiple of 3. It divides by 3.

28 548 is divisible by 6.

8 The last three digits are 548. You need to check if this divides by 8.

$$\begin{array}{r} 0\ 6\ 8 \\ 8\overline{)5\ ^54\ ^68} \end{array} \text{ remainder 4}$$

The last three digits do not divide by 8, so 28 548 is not divisible by 8.

9 $2 + 8 + 5 + 4 + 8 = 27$

27 is a multiple of 9. ● ─── You can repeat this if necessary, $2 + 7 = 9$

28 548 is divisible by 9.

10 The final digit is 8 so 28 548 is not divisible by 10.

28 548 is divisable by 2, 3, 4, 6, and 9. It is not divisable by 5, 8 or 10.

Exercise 1.1

1 Write down the inverse of each operation.

    a +10   **b** ×5   **c** −7   **d** +100   **e** ÷12

2 Write down the first six multiples of these numbers.

    a 5   **b** 9   **c** 3   **d** 10   **e** 7

3 Write down the 5th multiple of 6.

4 Write down the 9th multiple of 4.

5 Work out:

    a the 6th multiple of 23   **b** the 9th multiple of 47   **c** the 7th multiple of 84.

6 434 is a multiple of 7.

    a Is 435 a multiple of 7? Give a reason for your answer.

    b Is 441 a multiple of 7? Give a reason for your answer.

7 a Write down the first seven multiples of 4.   **b** Write down the first seven multiples of 6.

    c What is the lowest common multiple of 4 and 6?   **d** Is 48 a common multiple of 4 and 6?

8 Find the lowest common multiple of 5 and 7.

9 Find the lowest common multiple of 2 and 9.

10 Use the tests for divisibility to check whether these statements are true or false.

    a 522 is a multiple of 6.   **b** 735 is a multiple of 3.   **c** 957 is a multiple of 2.

    d 1550 is a multiple of 5.   **e** 4527 is a multiple of 9.   **f** 71 543 is a multiple of 5.

11 Ariana has a bag of sweets. She wants to share them with some friends.

She could share them between 8 people equally.

She could share them between 9 people equally.

She could share them between 12 people equally.

She knows she has fewer than 100 sweets.

How many sweets are there in the bag?

> Make a list of the numbers that divide by 8. Then do the same for 9 and 12.

12 Axel has a train set with two circular tracks.

He puts a fast train on one track and a slow train on the other track.

The fast train completes a circuit in 12 seconds.

The slow train completes a circuit in 15 seconds.

He starts them both at the station at the same time.

How many seconds before they are both at the station again at the same time?

> Write down how long each train takes to do 2 circuits, 3 circuits, etc.

13 There are several tests for divisibility by 7, but they are not easy to use. Here is one of them.

> Double the final digit. Subtract that from the number formed by the other digits.
>
> If the new answer divides by 7 then the original number will divide by 7 as well.

Here is an example using 483.

Double the final digit: $3 \times 2 = 6$

Subtract 6 from the number made by the other digits: $48 - 6 = 42$

42 is divisible by 7. So 483 is also divisible by 7.

Use this test to find out if these numbers are divisible by 7.

    a 581   **b** 518   **c** 408

1.2 Factors

Think of your multiplication tables.

Here are all the multiplications with 12 as the answer.

$1 \times 12 = 12$	$2 \times 6 = 12$	$3 \times 4 = 12$
or $\quad 12 \times 1 = 12$	$6 \times 2 = 12$	$4 \times 3 = 12$

The factors of 12 are 1, 2, 3, 4, 6 and 12.

The factors of a number divide into it with no remainder.

A factor must be an integer, which means it is a whole number.

You could write $1\frac{1}{2} \times 8 = 12$ but $1\frac{1}{2}$ is not a factor of 12 because it is not an integer.

Worked example 1

Find all the factors of 30.

$1 \times 30 = 30$ start with 1 because 1 is a factor of every number

$2 \times 15 = 30$ 30 is even so 2 is a factor

$3 \times 10 = 30$ try 3 next

$5 \times 6 = 30$ 4 is not a factor, so you try 5 next

The next number is 6 – but you already have that so you can stop here.

The factors of 30 are 1, 2, 3, 5, 6, 10, 15 and 30.

It helps if you list the factors in order although it is not necessary.

Common factors

You have just found the factors of 30. You can find the factors of 24 in the same way.

$1 \times 24 = 24$	$2 \times 12 = 24$	$3 \times 8 = 24$	$4 \times 6 = 24$

The factors of 24 are **1**, **2**, **3**, 4, **6**, 8, 12 and 24.

The factors of 30 are **1**, **2**, **3**, 5, **6**, 10, 15 and 30.

The numbers 1, 2, 3 and 6 are in both lists. They are common factors of 24 and 30.

The highest common factor of 24 and 30 is 6.

Worked example 2

Find the common factors of 28 and 42.

The factors of 28 are 1, 2, 4, 7, 14 and 28. •

Remember your multiplication tables:
$1 \times 28 = 28$ $2 \times 14 = 28$ $4 \times 7 = 28$

The factors of 42 are 1, 2, 3, 6, 7, 14, 21 and 42.

Remember your multiplication tables:
$1 \times 42 = 42$ $2 \times 21 = 42$
$3 \times 14 = 42$ $6 \times 7 = 42$

The common factors are 1, 2, 7 and 14.

Worked example 3

Use the tests for divisibility to find if 3 is a factor of 43 886.

Add the digits. $4 + 3 + 8 + 8 + 6 = 29$

29 is not a multiple of 3.

3 is not a factor of 43 886.

Exercise 1.2

1 Find all the factors of these numbers.

   a 12   **b** 8   **c** 15   **d** 9   **e** 17

   f 54   **g** 48   **h** 13   **i** 21   **j** 16

2 Find all the factors of 100.

3 a Write down all the factors of 18.

   b Write down all the factors of 24.

   c List all the common factors of 18 and 24.

   d What is the highest common factor of 18 and 24?

4 a Write down all the factors of 27.

   b Write down all the factors of 45.

   c List all the common factors of 27 and 45.

   d What is the highest common factor of 27 and 45?

5 Find all the common factors of 35 and 63.

6 Find all the factors of the following numbers.

   a 5   **b** 11   **c** 7   **d** 17   **e** 23   **f** 31

   g What do you notice about all of the answers?

7 a Find all the factors of 15.

   b The factors of 15 are all odd. Find another number that has no even factors.

   c Is it possible for a number to have no odd factors? Explain your answer.

8 a Jamie says that all numbers have at least two factors.
Is he right? Explain your answer.

   b Annie says that all numbers have an even number of factors.
Is she right? Explain your answer.

   c Jerome says that bigger numbers have more factors than smaller numbers. Is he right? Explain your answer.

> Look back at your answers to question **1** to help with this question.

1.3 Prime numbers

Look at the answers to question **6** in Exercise 1.2.

Numbers like these, with exactly two factors, are called **prime numbers**.

The factors of a prime number are 1 and the number itself.

The first few prime numbers are 2, 3, 5, 7, 11, 13, …

> Note that 1 is not a prime number. It has only one factor and prime numbers have **exactly** two factors.

Worked example

Which of the following are prime numbers?

a 17   **b** 21   **c** 27   **d** 41

a $1 \times 17 = 17$. There are no other factors of 17.

17 is a prime number.

b $1 \times 21 = 21$ and $3 \times 7 = 21$. There are four factors of 21.

21 is not prime.

c $1 \times 27 = 27$ and $3 \times 9 = 27$. There are four factors of 27.

27 is not prime.

d $1 \times 41 = 41$. There are no other factors of 41.

41 is a prime number.

> To check that a number is not prime think of the multiplication tables.
>
> If the number appears as an answer to more than one multiplication, then it is not a prime number.

The sieve of Eratosthenes

Eratosthenes was an ancient Greek mathematician. He was also an astronomer, poet and geographer.

He devised a simple method for checking prime numbers using a grid.

This grid goes up to 100 but it could go up to any number.

1	2	3	4	5	6	7	8	9	10
11	12	13	14	15	16	17	18	19	20
21	22	23	24	25	26	27	28	29	30
31	32	33	34	35	36	37	38	39	40
41	42	43	44	45	46	47	48	49	50
51	52	53	54	55	56	57	58	59	60
61	62	63	64	65	66	67	68	69	70
71	72	73	74	75	76	77	78	79	80
81	82	83	84	85	86	87	88	89	90
91	92	93	94	95	96	97	98	99	100

Copy the grid and follow these steps to find the prime numbers up to 100.

1 Cross out 1 because 1 is not a prime number.

2 Circle the next number that has not been crossed out, 2.

3 Cross out 4, 6, 8 and all the other multiples of 2.

4 Circle the next number that has not been crossed out, 3.

5 Cross out 9, 15, 21 and all the other multiples of 3 that have not already been crossed out.

The grid should now look like this.

1	2	3	4	5	6	7	8	9	10
11	12	13	14	15	16	17	18	19	20
21	22	23	24	25	26	27	28	29	30
31	32	33	34	35	36	37	38	39	40
41	42	43	44	45	46	47	48	49	50
51	52	53	54	55	56	57	58	59	60
61	62	63	64	65	66	67	68	69	70
71	72	73	74	75	76	77	78	79	80
81	82	83	84	85	86	87	88	89	90
91	92	93	94	95	96	97	98	99	100

Now follow these steps.

6 Circle the next number that has not been crossed out.

7 Cross out all the other multiples of that number that have not already gone.

Repeat steps **6** and **7** until all you have left is a set of circled numbers.

They are the prime numbers less than 100.

Exercise 1.3

Use your prime number grid to answer these questions.

1 How many prime numbers are there in the second row of the grid from 11 to 20?

2 How many prime numbers are there from 21 to 30?

3 How many prime numbers less than 100 end in the digit 1?

4 How many prime numbers less than 100 end in the digit 7?

5 There are six digits that prime numbers can never end in. Write down all of them.

6 Use the tests for divisibility to check whether these are prime numbers.
For those that are not prime, write down what number is a factor.
For example, 125 is not prime because 5 is a factor.

    a 106   **b** 107   **c** 111   **d** 123   **e** 127

7 An example of a run of non-prime numbers is 8, 9, 10. This is a run of three.

What is the longest non-prime run you can find less than 100?

8 20 can be written as the sum of two primes. For example, 20 = 13 + 7

> Use your grid of prime numbers to help you answer questions **7** and **8**.

    a Can 20 be written as the sum of two primes in any other way?

    b Write these numbers as the sum of two primes in as many ways as possible.

      i 21   **ii** 22   **iii** 24   **iv** 30   **v** 70

    c Can you find any numbers that cannot be written this way?

9 Matt claims that he has found a sequence of prime numbers starting with 11 as follows:

$$11 \xrightarrow{+2} 13 \xrightarrow{+4} 17 \xrightarrow{+6} 23 \xrightarrow{+8} 31, \text{ etc.}$$

Investigate to see whether this sequence continues to give prime numbers.

> Continue the number pattern and use your grid of prime numbers to check.

10 Look at your answers to questions **1** and **2**.

Between 11 and 20 there are four prime numbers.

Between 21 and 30 there are two prime numbers.

    a Look at some more rows in the prime number grid.

      What is the highest number of prime numbers that appears in any row?

    b What is the lowest number of prime numbers that appears in any row?

11 a The factors of 6 are 1, 2, 3 and 6.

      Ignore the number 6 itself and add together the other factors.

      Write down the answer.

In part **a** the other factors of 6 add up to 6.

A number like this, where the factors add up to the number, is called a **perfect number**.

    b Write down all of the factors of 28 which are less than 28.

      Add them together and show that 28 is also a perfect number.

    c Perfect numbers are rare. The next one after 28 is 496.

      Find all the factors of 496 and show that it is a perfect number. (Try 1, then 2, then 3, etc.)

1.4 Squares and square roots

Squares

Look at this sequence of shapes made up of small tiles.

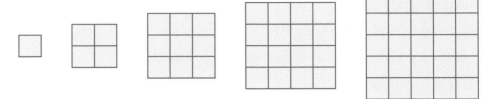

The first shape is 1 by 1.   It has 1 tile.

The second shape is 2 by 2.   It has 4 tiles.

The third shape is 3 by 3.   It has 9 tiles.

The fourth shape is 4 by 4.   It has 16 tiles.

The fifth shape is 5 by 5.   It has 25 tiles.

Without a diagram you can see that the next shape would be 6 by 6.

It would have $6 \times 6 = 36$ tiles.

The numbers 1, 4, 9, 16, 25, 36, ... are called **square numbers**.

> They are called square numbers because the shapes they make are square – not because of the square tiles.

You can write the square numbers like this: $1^2 = 1$. This is read as 1 squared equals 1.

The sequence continues $2^2 = 4$

$3^2 = 9$

$4^2 = 16$

$5^2 = 25$, etc.

You need to learn all the square numbers from 1 to 400.

Worked example

a What is the 9th square number?

b Is 72 a square number?

a The 9th square number is $9 \times 9 = 81$

b $8 \times 8 = 64$ 64 is a square number

$9 \times 9 = 81$ 81 is a square number

There is no square number between 64 and 81.

72 is not a square number.

Square roots

You know that $6^2 = 36$. You can also write that the **square root** of 36 is 6, or $\sqrt{36} = 6$

You also know that $\sqrt{1} = 1$, $\sqrt{4} = 2$, $\sqrt{9} = 3$, etc.

Square root and square are inverses of each other.

If $20^2 = 400$ then $\sqrt{400} = 20$

You may sometimes need to use a calculator for squares and square roots.

Most calculators have a key like this $\boxed{x^2}$ for squares and one like this $\boxed{\sqrt{x}}$ for square roots.

Exercise 1.4

1 Write down the 8th square number.

2 Work out and write down all of the square numbers from 1^2 to 10^2.

3 Work out the square numbers between 100 and 200.

4 Find: **a** 16^2 **b** 18^2 **c** 20^2 **d** 25^2 **e** 30^2

5 You can write 20 as the sum of two square numbers: $20 = 16 + 4$

Write these numbers as the sum of two square numbers.

a 13   **b** 25   **c** 29   **d** 45   **e** 100

f Find a number that cannot be written in this way.

6 Write down the value of:

a $\sqrt{100}$ **b** $\sqrt{64}$ **c** $\sqrt{121}$ **d** $\sqrt{49}$

7 Find the value of:

a $\sqrt{196}$ **b** $\sqrt{361}$ **c** $\sqrt{441}$ **d** $\sqrt{676}$

8 Use your answers to question **2** to answer this question.

a Khalil says that 317 cannot be a square number because it ends in 7. Is he right?

b Mahri says that square numbers never end in 3. Is she right?

> Look at the final digit of each of the square numbers in questions **2**, **3** and **4**.

c Aziz says there are four digits that square numbers never end in. Is he right? If so, what are the four digits?

1.5 Negative numbers

Here is a **number line** with positive and negative integers shown.

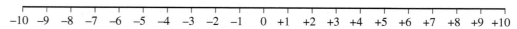

So far in this chapter you have only used **positive** numbers but now you will learn to work with **negative** numbers as well. Together, these numbers are called **directed numbers**.

Negative numbers are often seen on a thermometer where negative temperatures are common.

A thermometer is a type of practical number line.

> Positive numbers can be written with a + sign but that is not necessary. For example, +3 is usually written as 3.

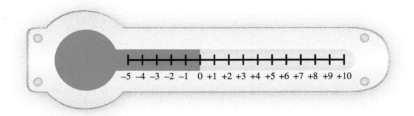

Worked example 1

Put these numbers in order, starting with the smallest.

$$5, \quad -6, \quad 2, \quad 8, \quad 0, \quad -1, \quad -8$$

You know that this is the same as +5, −6, +2, +8, 0, −1, −8.

You can see from the number line that the correct order is −8, −6, −1, 0, 2, 5, 8.

Adding and subtracting directed numbers

You can add directed numbers by moving along the number line.

To add a positive number you move towards the right.

To add a negative number you move towards the left.

You can subtract directed numbers by moving along the number line in the opposite direction.

To subtract a positive number you move towards the left.

To subtract a negative number you move towards the right.

Worked example 2

Use the number line to work out:

a $(-4) + (+7)$ **b** $(-1) + (-5)$ **c** $(-2) - (+4)$ **d** $3 - (-5)$

a To add a positive number you need to move to the right.

Start at (-4) on the number line and count 7 steps to the right.

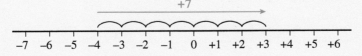

The answer is +3. This could be written as just 3.

b To add a negative number you need to move to the left.

Start at (-1) on the number line and count 5 steps to the left.

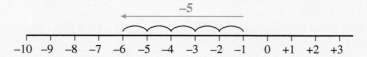

The answer is -6.

c To subtract a positive number you need to move to the left.

Start at (-2) on the number line and count 4 steps to the left.

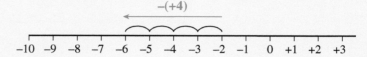

The answer is -6.

d To subtract a negative number you need to move to the right.

Start at 3 on the number line and count 5 steps to the right.

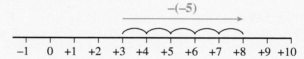

The answer is $+8$ or 8.

Exercise 1.5

1 Write these numbers in order, starting with the smallest.

    a $6, 1, -7, -9, -2, 5$ **b** $-10, 3, 8, 0, -3, -7$

    c $12, 5, -4, 9, -14, -20$ **d** $-23, -13, 7, 0, -9, 16$

2 The daytime temperature in Kiev is $5\,°C$. At night it is $8\,°C$ colder.

    What is the night-time temperature?

3 Copy and complete this table. The first row has been done for you.

Daytime temperature	Night-time temperature	Change in temperature
$6\,°C$	$-1\,°C$	Down $7\,°C$
$9\,°C$	$-4\,°C$	Down 13 °C
$21\,°C$	$8\,°C$	Down 13 °C
$12\,°C$	-4 °C	Down $16\,°C$
$4\,°C$	-5 °C	Down $9\,°C$
5 °C	$-2\,°C$	Down $7\,°C$
11 °C	$-12\,°C$	Down $23\,°C$

For question **4** onwards, use the number line to work out:

4 $(-3) + (+6)$ **5** $(-8) + (+5)$ **6** $0 + (+7)$ **7** $(-10) + (+14)$

8 $(-2) + (-4)$ **9** $5 + (-7)$ **10** $6 + (-6)$ **11** $(-4) + (-5)$

12 $(-5) - (+4)$ **13** $6 - (+8)$ **14** $(-1) - (+5)$ **15** $9 - (+12)$

16 $(-3) - (-7)$ **17** $5 - (-2)$ **18** $(-6) - (-6)$ **19** $(-1) - (-10)$

20 $(-4) + (-6)$ **21** $3 - (+5)$ **22** $(-2) - (-9)$ **23** $(-2) + (+12)$

1.6 More adding and subtracting

When adding or subtracting larger numbers it is not easy to use the number line.

The rules for combining the signs are:

Change + + to +   e.g. $-4 + (+6)$ becomes $-4 + 6 = 2$

Change + − to −   e.g. $-4 + (-6)$ becomes $-4 - 6 = -10$

Change − + to −   e.g. $-4 - (+6)$ becomes $-4 - 6 = -10$

Change − − to +   e.g. $-4 - (-6)$ becomes $-4 + 6 = 2$

Worked example

Use the rules for combining signs to work out:

a $(-8) + (+11)$ **b** $(-6) + (-12)$ **c** $(-6) - (+14)$ **d** $13 - (-20)$

a $(-8) + (+11)$ becomes $-8 + 11 = 3$ **b** $(-6) + (-12)$ becomes $-6 - 12 = -18$

c $(-6) - (+14)$ becomes $-6 - 14 = -20$ **d** $13 - (-20)$ becomes $13 + 20 = 33$

Exercise 1.6

Work out these.

1 $(-6) + (+7)$ **2** $4 + (-9)$ **3** $(-1) - (+6)$ **4** $3 - (-13)$

5 $(-5) + (-6)$ **6** $(-1) + (+9)$ **7** $(-4) - (-8)$ **8** $0 + (+6)$

9 $(-13) + (+17)$ **10** $(-14) + (-13)$ **11** $9 - (+2)$ **12** $0 - (+32)$

13 $21 + (-15)$ **14** $(-8) - (+15)$ **15** $(-16) - (-23)$ **16** $(-14) + (-16)$

17 $(-11) - (-18)$ **18** $(-4) - (-23)$ **19** $(-5) + (-21)$ **20** $(-21) + (+19)$

21 $13 - (+8)$ **22** $15 - (+18)$ **23** $12 - (-8)$ **24** $(-56) + (+32)$

1.7 Further calculations

Order of operations

In a calculation like $2 + 3 \times 5$ the order in which you work it out is important.

If the addition is done first, then $2 + 3 \times 5 = 5 \times 5 = 25$

If the multiplication comes first, then $2 + 3 \times 5 = 2 + 15 = 17$

To avoid any confusion there is a correct order to do the operations.

If there are any **brackets** – work them out first.

If there are any powers or **indices**, they come second.

Division and **multiplication** are next in order.

Finally come **addition** and **subtraction**.

The initials of the words in bold spell out **BIDMAS**.

1 Brackets

2 Indices or powers

3 Division
4 Multiplication

5 Addition
6 Subtraction

Note the spacing – division and multiplication have equal importance.

Addition and subtraction have equal importance.

Worked example 1

Work out the answers to these.

a $6 + 12 \div 2$ **b** $16 \div (-4) \times 2$ **c** $(24 - 9) + 4 \times -3$

a $6 + 12 \div 2$ division before addition

$= 6 + 6$ addition next

$= 12$

b $16 \div (-4) \times 2$ division and multiplication are equal so start at the left

$= -4 \times 2$

$= -8$

c $(24 - 9) + 4 \times -3$ work out the brackets first

$= 15 + 4 \times -3$ multiplication before addition

$= 15 + -12$

$= 3$

Mental calculations

When you are doing a calculation in your head it is sometimes possible to find a shortcut.

For example, if you have this calculation:

$17 + 31 - 5 + 9 + 25 + 13$

Look for pairs of numbers that add or subtract to give a round number.

In this case, $31 + 9 = 40$ and $17 + 13 = 30$ and $-5 + 25 = 20$

Re-arrange the calculation to group these together, as follows:

$17 + 31 - 5 + 9 + 25 + 13$

$= 17 + 13 + 31 + 9 - 5 + 25$

$= 30 + 40 + 20$

$= 90$

> You can re-arrange this because addition and subtraction are paired together in the BIDMAS list above.

These examples show you some more techniques to look out for.

Worked example 2

Work out the answers to these mentally.

a $3 \times 56 + 7 \times 56$ **b** $16 \times 12 \div 8$ **c** $(24 - 9) \times 5 \div 3$

a $3 \times 56 + 7 \times 56$
In this question there are 3 'fifty sixes' and 7 'fifty sixes'.
Altogether there are 10 'fifty sixes'.
You can think of it as follows:
$3 \times 56 + 7 \times 56$
$= 10 \times 56$
$= 560$

b $16 \times 12 \div 8$
This calculation can be simplified by re-arrangement.
$16 \times 12 \div 8$
$= 16 \div 8 \times 12$
$= 2 \times 12$
$= 24$

> You can re-arrange this because multiplication and division are paired together in the BIDMAS list above.

c $(24 - 9) \times 9 \div 3$
$= 15 \times 9 \div 3$ work out the bracket first
$= 15 \div 3 \times 9$ re-arrange to do $15 \div 3$ next
$= 5 \times 9$
$= 45$

Exercise 1.7

Work out these.

1 $5 + 4 \times 2$ **2** $4 + 7 - 5$ **3** $6 \times 4 \div 8$ **4** $3 \times 3 - 7$

5 $(9 - 5) \times 3$ **6** $(2 + 5) \times (6 - 1)$ **7** $20 - 12 \div 4$ **8** $18 - 9 - 5$

9 $18 - (9 - 5)$ **10** $8 + 9 \div 3$ **11** $45 - 20 + 7$ **12** $45 - (20 + 7)$

13 $48 \div 12 \times 2$ **14** $9 + 4 \times 3 - 6$ **15** $25 - 3 + 2 \times 2$ **16** $16 \div 4 + 3 \times 5$

Work out the answers to questions **17** to **24** in your head.

17 $36 + 18 - 16 + 12$ **18** $23 + 45 - 13 - 15$ **19** $12 \times 82 - 2 \times 82$

20 $6 \times 164 + 4 \times 164$ **21** $24 \times 15 \div 8$ **22** $72 \times 3 \div 9$

23 $9 \times 13 + 4 \times 13 - 3 \times 13$ **24** $(7 + 13) \times 18 \div 10$

25 Put brackets in the right place to make these calculations correct.

    a $8 + 6 \div 2 = 7$ **b** $16 - 4 + 5 = 7$

    c $6 \times 2 + 5 = 42$ **d** $2 + 3 \times 8 - 5 = 15$

> Write them out several times and try putting brackets in different places.

26 Work out these.

    a $(-2) \times (-6) + 9$ **b** $3 + 5^2$ **c** $(-9) \div 3 + (-4)$ **d** $12 - (-5) \times (-4)$

    e $6 \times 4 - 2^3$ **f** $6^2 - 4^2 \div 8$ **g** $19 + (-20) \div (-5)$ **h** $(-6) + (-8) \div (-2) - (-7)$

> Use BIDMAS and the rules for directed numbers.

Review

1 Write down the inverse of these operations.

    a $\times 9$ **b** $\div 16$ **c** -45 **d** square   **e** $+200$ **f** square root

2 Write down the first five multiples of these numbers.

    a 4   **b** 11   **c** 20   **d** 15

3 Write down the first six multiples of:

    a 12   **b** 9   **c** What is the lowest common multiple of 12 and 9?

    d Is 108 also a common multiple of 12 and 9?

4 a Ravi says that 882 is a common multiple of 9 and 12. Is he right?

    b He also thinks that 882 is a multiple of 7. Use division to check whether he is correct.

    c What is the smallest prime number which is not a factor of 882?

5 Find all of the factors of these numbers.

    a 7   **b** 10   **c** 18

    d 29   **e** 36   **f** 25

6 Find all the factors of:

    a 28   **b** 84   **c** What is the highest common factor of 28 and 84?

7 Find all the factors of the following numbers.

    a 4   **b** 16   **c** 25   **d** 49   **e** 100

    f What do you notice about the number of factors in each case?

    g What type of numbers are these?

8 Which of the following are prime numbers?

    a 13   **b** 23   **c** 33   **d** 19   **e** 69

9 Use tests for divisibility to check which of these are prime numbers.

    a 136   **b** 137   **c** 147   **d** 187   **e** 191

10 Write down the value of:

    a 6^2 **b** 4^2 **c** 11^2 **d** 12^2 **e** 20^2

11 Work out:

    a 26^2 **b** 19^2 **c** 23^2 **d** 18^2 **e** 40^2

12 Write down:

    a $\sqrt{49}$ **b** $\sqrt{81}$ **c** $\sqrt{121}$ **d** $\sqrt{400}$ **e** $\sqrt{196}$

13 Find the value of:

    a $\sqrt{576}$ **b** $\sqrt{1225}$ **c** $\sqrt{169}$ **d** $\sqrt{10\,000}$

14 Write these numbers in order, starting with the smallest:

    a $13, 8, -14, -5, -23$ **b** $0, -34, 24, 18, -18, 2$

15 Work out these.

    a $5 + 9$ **b** $(-9) + 4$ **c** $8 + (-7)$ **d** $(-15) + (-5)$

    e $(-7) + (-4)$ **f** $(-18) + (+6)$ **g** $0 + (-19)$ **h** $(-21) + (+21)$

    i $12 - 7$ **j** $(+13) - (+24)$ **k** $(+8) - (-6)$ **l** $(-7) - (+14)$

    m $(-25) - (-13)$ **n** $(-42) - (-6)$ **o** $(-20) - (-20)$

16 Work out these.

    a $42 \div (9 - 2)$ **b** $12 + 8 \div 4$ **c** $24 - 32 \div 8$

    d $6 \times (20 - 17)$ **e** $18 \div 3 \times 5$ **f** $4 + 12 \div (4 - 2)$

    g $18 - (-4) + 12 \div (-6)$ **h** $(-3) \times (5 + (-4))$

17 Here are statements about directed numbers.

Some of the statements are always true.

Some of the statements are always false.

Some of the statements are sometimes true.

> Try using several different pairs of numbers to see what happens.

For each statement decide whether it is 'always true', 'always false' or 'sometimes true'.

    a When you add two positive numbers the answer is positive.

    b When you add two negative numbers the answer is positive.

    c When you add a negative number to a positive number the answer is negative.

    d When you subtract a positive number from a positive number the answer is positive.

    e When you subtract a negative number from a positive number the answer is positive.

    f When you subtract a negative number from a negative number the answer is negative.

2 Expressions

Learning outcomes

- Construct simple algebraic expressions by using letters to represent numbers.
- Simplify linear expressions, e.g. collect like terms; multiply a constant over a bracket.
- Substitute positive integers into simple linear expressions.

Algebra

Mathematicians use letters to represent unknown numbers. Using letters to represent unknown numbers is called algebra.

Algebra is often called the language of mathematics. It is a language that is hundreds of years old and was first developed by the Babylonians.

Algebra is a language that is understood in every country around the world.

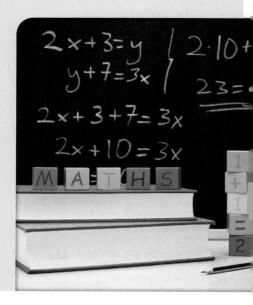

2.1 Using letters for unknown numbers

Solving problems is often much easier if you use letters to represent unknown numbers.

Consider the following problem:

A bag contains an unknown number of marbles.

Two more marbles are added to the bag.

How many marbles are there in the bag now?

Choose a letter to represent the unknown number of marbles.

Let n = the unknown number of marbles.

n is called a variable.

Two more marbles are added to the bag.

There are now $n + 2$ marbles in the bag.

$n + 2$ is called an expression.

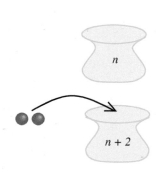

Some rules of algebra:

- $x + x$ means 2 lots of x so $x + x = 2 \times x = 2x$ ●——— The number is always written before the letter.

- $y \times x$ is the same as $x \times y$ and is written as xy ●—(Put the letters in alphabetical order.)

- $7 \times y \times x = 7xy$

- $x \div 2$ is written as $\frac{x}{2}$

Worked example 1

Write this sentence as an expression:

Start with the number n, multiply by 2 and then add 5.

Build up the expression in stages:

Start:	n
Multiply by 2:	$2n$
Add 5:	$2n + 5$

Worked example 2

A melon costs m cents and a peach costs p cents.

Write down an expression for the cost in cents of:

a 7 melons   **b** 5 peaches   **c** 7 melons and 5 peaches.

a Cost of 7 melons = $7 \times m = 7m$

b Cost of 5 peaches = $5 \times p = 5p$

c Cost of 7 melons and 5 peaches = $7m + 5p$

Exercise 2.1

1 Write these sentences as expressions.

     a Start with the number p and then multiply by 3.

     b Start with the number A and then add 4.

     c Start with the number d and then take away 8.

     d Start with the number b and then multiply by 5.

     e Start with the number x, multiply by 4 and then add 3.

     f Start with the number y, multiply by 2 and then take away 5.

     g Start with the number Q, multiply by 7 and then add 1.

h Start with the number R and then take it away from 8.

i Start with the number c, multiply by 5 and then add d.

j Start with the number f, add g and then add 2.

k Start with the number j, take away k and then add 2.

l Start with the number m, multiply it by 5 and then take away from 2.

2 There are x fish in a pond.

Omar puts 5 more fish in the pond.

How many fish are in the pond now?

3 A piece of string is y cm long.

Farrida cuts off 3 cm.

How long is the piece of string now?

4 A brick has a mass of m kg.

Write an expression for the mass of 4 bricks.

5 Gregor was h cm tall.

His height increases by 2 cm.

Write an expression for his new height.

6 Raju is x years old.

a How old was Raju 4 years ago?

b How old will Raju be 6 years from now?

7 Robert, Anna and Helen have some marbles.

Anna has 5 marbles more than Robert.

Helen has 3 marbles fewer than Robert.

If Anna has x marbles:

a how many marbles does Robert have?

b how many marbles does Helen have?

8 A pencil costs p cents and a ruler costs r cents.

Write an expression for the cost of:

a 5 pencils　　**b** 2 rulers　　**c** 5 pencils and 2 rulers.

9 A cup of coffee costs c dollars and a cup of tea costs t dollars.

Write an expression for the cost of:

a 3 cups of coffee　　**b** 7 cups of tea　　**c** 3 cups of coffee and 7 cups of tea.

10 A CD costs x dollars and a DVD costs y dollars.

Write an expression for the cost of 3 CDs and 5 DVDs.

11 Write an expression for the perimeter of each of these shapes.

a

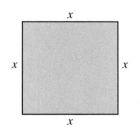

b

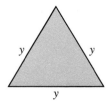

c

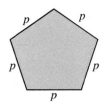

12 Match each statement with the correct expression.

The first one has been done for you.

Subtract 2 from x and then multiply by 5.

Add 2 to x and then multiply by 5.

Multiply x by 2 and then subtract from 5.

Multiply x by 5 and then subtract 2.

Multiply x by 2 and then add 5.

$5(x + 2)$

$5 - 2x$

$5x - 2$

$2x + 5$

$5(x - 2)$

> Note that
> $5(x + 2)$ means
> $5 \times (x + 2)$

13 The cost of a theatre ticket for an adult is $\$a$.

The cost of a theatre ticket for a child is $\$c$.

Adult tickets are more expensive than child tickets.

a Match each statement with the correct expression.

The total cost in dollars of 5 adult tickets and 4 child tickets

The difference in price in dollars between an adult ticket and a child ticket

The cost in dollars of 5 child tickets

The change in dollars from $100 when you buy 4 adult tickets

$4a + 5c$

$100 - 4a$

$5c$

$5a + 4c$

$a - c$

$100 - 9c$

b Write a statement for each of the two expressions that are left.

2.2 Simplifying expressions

The expression $5x$ has one **term**.

The expression $2x + 5y - 3z + 7$ has four terms.

The expression $7x + 8y + 5x$ has two **like terms** that can be collected together.

> Like terms contain the same letter.

The like terms are the $7x$ and the $5x$.

$$7x + 8y + 5x$$
$$= 7x + 5x + 8y$$
$$= 12x + 8y$$

> Note that the sign in front of each term stays with the term.

Worked example

Simplify these expressions.

a $5x + 3x$ **b** $7m - m$ **c** $2x + 8y + 4x - 5y$ **d** $8x + 3 - 4x - 10$

a $5x + 3x = 8x$

b $7m - m = 7m - 1m = 6m$ m is the same as $1m$

c $2x + 8y + 4x - 5y = 2x + 8y + 4x - 5y$ find the like terms
$$= 2x + 4x + 8y - 5y \quad \text{put the like terms next to each other}$$
$$= 6x + 3y$$

d $8x + 3 - 4x - 10 = 8x + 3 - 4x - 10$ find the like terms
$$= 8x - 4x + 3 - 10 \quad \text{put the like terms next to each other}$$
$$= 4x - 7$$

Exercise 2.2

1 Simplify these expressions.

  a $y + y + y$ **b** $5x + 2x$ **c** $7p - 2p$ **d** $8a + a$

  e $5q - q$ **f** $14m - 2m$ **g** $15y + 4y$ **h** $5h - 6h$

  i $5x + 2x + 4x$ **j** $7y - 3y + 2y$ **k** $8p - 2p - 3p$ **l** $5p + q - 4q$

  m $7h - 2h - 8h$ **n** $5p - 8p + 10p$ **o** $8q - q - 7q$ **p** $-8f + 2f + 2f$

2 This is an algebraic wall.

To find the expression in each block you add the expressions in the two blocks below it.

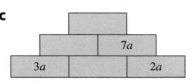

Copy and complete these algebraic walls.

a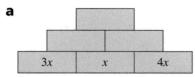

b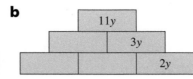

c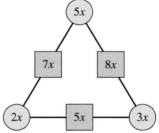

3 To find the expression in each square you add the expressions in the two connecting circles.

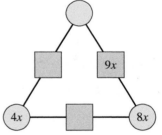

Copy and complete these diagrams.

a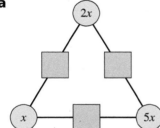

b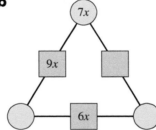

c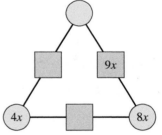

4 Hassan has got his simplifying expressions homework wrong.

Copy out each question and correct his mistakes.

a $8y - 3x + 2y$
$= 8y - 2y + 3x$
$= 6y + 3x$

b $5 + 3x - 2 + 6x$
$= 5 - 2 + 3x + 6x$
$= 3 + 9x$
$= 12x$

c $4xy + 8x + 2yx + 7x$
$= 15x + 4xy + 2yx$

5 Simplify these expressions.

a $8x + 3y + 4x + 5y$ **b** $9f + 4g + 6f + g$ **c** $7x + 5 + 2x + 7$

d $6p + 4q - 2p - 2q$ **e** $8x + 7 - x - 4$ **f** $5a - 2b + 4a + 8b$

g $9f - 8g + 3f + 2g$ **h** $8x + 10y - 7x + 2y$ **i** $-2x - 3y + 7x + 8y$

j $3y + 5x - 8 - 2y$ **k** $5m - 8 + 2m - 1$ **l** $8 + 2y - 5x + 3y$

6 Copy and complete.

a $5x + \square + 3x + 4y = 8x + 11y$ **b** $\square + 2x + 7 + 4x = 10 + 6x$

c $6x - 2y + \square - 4y = 9x - 6y$ **d** $8p + 7q + \square - \square = 11p + 5q$

e $9 + 3h + 4 - \square = 13 + 7h$ **f** $14 - 5x + 2y - \square + \square = 14 - 7x + 8y$

7 Copy and complete these algebraic walls.

a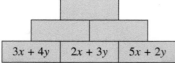

3x + 4y | 2x + 3y | 5x + 2y

b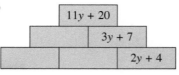

11y + 20
3y + 7
2y + 4

c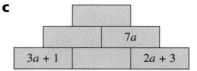

7a
3a + 1 | 2a + 3

8 Write an expression for the perimeter of each of these shapes.

Write your answers in their simplest form.

a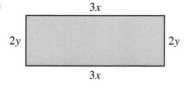

3x, 2y, 2y, 3x

b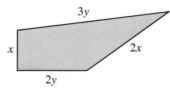

3y, x, 2x, 2y

c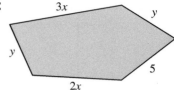

3x, y, y, 5, 2x

9 Simplify these expressions.

a $5x + 3xy + 2x + 8xy$

b $7ab + 3ba + 5a$

c $7 + 3xy + 8yx$

d $6fg + 2g + 5gf$

e $7mn + 2 + 3nm$

f $8xy + 7 + 2x - 3yx$

g $8y - 9xy - 4y$

h $6fg - 4ad - 3gf - da$

i $3pq - 2qr + 4rq - 2qp$

10 The diagram shows some expressions that are equivalent to $5x + 2y$

Copy the diagram and find four more expressions that are equivalent to $5x + 2y$

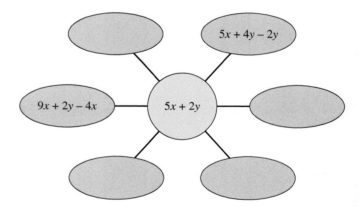

5x + 4y − 2y

9x + 2y − 4x

5x + 2y

11 Copy and complete the algebraic wall.

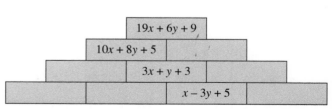

19x + 6y + 9
10x + 8y + 5
3x + y + 3
x − 3y + 5

12 Look at your diagrams for question **3**.

Can you find a rule for finding the expressions in each of the circles when you know the expressions in the squares?

Write the rule for finding the expression in the orange circle using the letters a, b and c.

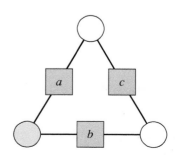

a, c, b

2.3 Expanding brackets

Some algebraic expressions contain brackets.

$3(x + 2)$ means 3 lots of $(x + 2)$

So $3(x + 2) = x + 2 + x + 2 + x + 2 = 3x + 6$

It is much quicker to multiply out the brackets.

When you multiply out the brackets you must multiply each term in the brackets by the term outside the brackets.

$$3(x + 2) = 3 \times x + 3 \times 2 = 3x + 6$$

$$7(2x + 3) = 7 \times 2x + 7 \times 3 = 14x + 21$$

Multiplying out the brackets is also called expanding brackets.

Worked example 1

Multiply out the brackets.

a $6(x + 7)$ **b** $3(y - 2)$ **c** $5(2 - 3x)$

a $6(x + 7) = 6 \times x + 6 \times 7$ multiply each term in the bracket by 6
$$= 6x + 42$$

b $3(y - 2) = 3 \times y - 3 \times 2$ multiply each term in the bracket by 3
$$= 3y - 6$$

c $5(2 - 3x) = 5 \times 2 - 5 \times 3x$ multiply each term in the bracket by 5
$$= 10 - 15x$$

Worked example 2

Expand and simplify: $7(x + 5) + 6(2x - 3)$

$$7(x + 5) + 6(2x - 3) = 7 \times x + 7 \times 5 + 6 \times 2x - 6 \times 3$$ expand each set of brackets
$$= 7x + 35 + 12x - 18$$ collect like terms
$$= 19x + 17$$

Exercise 2.3

1 Nadia has got her expanding brackets homework wrong.

Copy out each question and correct her mistakes.

a $2(x + 1)$
 $= 2x + 3$

b $5(2 + 3x)$
 $= 5 \times 5x$
 $= 25x$

c $7(2x - 8)$
 $= 14x - 8$

2 Multiply out the brackets.

a $3(x + 4)$ **b** $5(y + 2)$ **c** $2(m + 7)$ **d** $4(p + 6)$

e $8(3 + x)$ **f** $9(2 + g)$ **g** $6(9 + x)$ **h** $7(8 + k)$

i $4(x - 3)$ **j** $5(y - 3)$ **k** $7(4 - y)$ **l** $6(15 - g)$

3 Expand the brackets.

a $3(2x + 7)$ **b** $4(3y + 8)$ **c** $5(2m + 1)$ **d** $4(7p + 2)$

e $6(3 + 7x)$ **f** $4(7 + 8g)$ **g** $3(9 + 5x)$ **h** $9(8 + 2k)$

i $4(9x - 2)$ **j** $6(8p - 3)$ **k** $3(14 - 5y)$ **l** $7(25 - 4g)$

4 Multiply out the brackets.

a $2(3x + 4y + 6)$ **b** $5(7x + 3y - 2)$ **c** $2(5m + 7n + 3)$

d $4(8p - 6q + 5)$ **e** $8(8x - 3 + 2y)$ **f** $9(3f + 2g - 7h)$

g $8(5x - 2y - 7)$ **h** $7(8p - 2q + 4r)$ **i** $4(9m + 3n - 4p)$

5 Multiply out the brackets and simplify.

a $3(x + 4) + 2(x + 5)$ **b** $5(y + 3) + 4(y + 1)$ **c** $4(m + 6) + 3(m + 7)$

d $4(p + 6) + 2(p + 5)$ **e** $7(2x + 3) + 4(3x + 5)$ **f** $7(2x + 9) + 4(5x + 2)$

g $6(9y + 4) + 3(5 + 2y)$ **h** $4(8 + 7k) + 3(6k + 7)$ **i** $4(2x + 3) + 2(3x - 5)$

j $5(4y - 3) + 4(2y + 7)$ **k** $9(4x - 2) + 7(3x + 5)$ **l** $6(3 - 2y) + 8(5y - 1)$

m $4(5x - 8) + 3(5x - 4)$ **n** $2(3 - 4g) + 5(3g + 2)$ **o** $8(5 - 2x) + 2(7 - 3x)$

p $4(2 + 5x) + 5(2 - 7x)$ **q** $3(7 + 8x) + 4(2 - 6x)$ **r** $4(7 - 3x) + 5(2x - 6)$

6 Multiply out the brackets and simplify.

a $3(x + 5) + 2$ **b** $5 + 7(y + 3)$ **c** $4(m + 6) + 8m$

d $4 + 8(p + 6) + 2$ **e** $7x + 3(2x + 4) + 9$ **f** $7 + 8(2x - 9) + 4$

g $6y + 2(9y + 4) - 8 + 3y$ **h** $4(3x + 2y) + 5(2y - 9x)$ **i** $4(5x + 2y) + 4(2x - 5y)$

7

| **A** $2x - 3$ | **B** $5 - x$ | **C** $8x - 5$ |

Expand and simplify.

a $5A + 3B$
b $2A + 3C$
c $3A + 2B + 4C$
d $7A + 2B + 5C - 8$

8 Copy and complete.

a $5(2x + \square) = 10x + 15$
b $3(7x - \square) = 21x - 12$

c $8(3 - \square x) = 24 - 16x$
d $8(2y - \square) = 16y - 72$

9 Find the two missing numbers:

$3(2x + \square) + \square(x - 2) = 11x + 14$

10 Here is a set of instructions.

Start with a number $\rightarrow$ add 3 $\rightarrow$ multiply by 2 $\rightarrow$ subtract 6 $\rightarrow$ divide by 2 $\rightarrow$ write down the answer.

So if you start with the number 9 you get : 9 $\rightarrow$ 12 $\rightarrow$ 24 $\rightarrow$ 18 $\rightarrow$ 9

a Using the instructions above copy and complete the following.

i 6 $\rightarrow \square \rightarrow \square \rightarrow \square \rightarrow \square$

ii 13 $\rightarrow \square \rightarrow \square \rightarrow \square \rightarrow \square$

iii 25 $\rightarrow \square \rightarrow \square \rightarrow \square \rightarrow \square$

b What do you notice about your answers in part **a**?

c Use algebra to explain your results to part **a**.

> Copy and complete
> $x \rightarrow \square \rightarrow \square \rightarrow \square \rightarrow \square$

2.4 Substitution into an expression

Substitution into an expression means replacing the letters in an expression by the given numbers.

Worked example

If $a = 3$, $b = 4$ and $c = 5$ find the value of these expressions.

a $7a + c$
b $4 + 2c$
c $ab - bc$

a $7a + c$

$= (7 \times 3) + 5$

$= 21 + 5$

$= 26$

b $4 + 2c$

$= 4 + (2 \times 5)$

$= 4 + 10$

$= 14$

c $ab - bc$

$= (3 \times 4) - (4 \times 5)$

$= 12 - 20$

$= -8$

Exercise 2.4

1 If $a = 6$ and $b = 7$ find the value of:

a $a + 4$ **b** $a - b$ **c** $a - 2$ **d** $5a$

e $2a + 3b$ **f** $5b - 2a$ **g** $4 + 3b$ **h** $3ab$

i $4(a + b)$ **j** $a(2 + b)$ **k** $\frac{a}{3}$ **l** $\frac{a}{2} + b$

m $\frac{12}{a} + 3$ **n** $\frac{a + 2b}{2}$ **o** $\frac{2b + 4}{a}$ **p** $\frac{2ab}{a}$

2 If $x = 4$ and $y = 5$ find the value of:

a $x + y$ **b** $2y$ **c** $x - y$ **d** $y - 2x$

e $\frac{12}{x}$ **f** $2 - \frac{x}{2}$ **g** $\frac{2x + 2y}{3}$ **h** $\frac{20}{x} + y$

3 If $p = 3$, $q = 4$ and $r = 2$ find the value of:

a $p + q + r$ **b** $p + q - r$ **c** $2p + 3q + 4r$ **d** $3p - 4q - 2r$

e $5(p + q - 2)$ **f** $pq + pr + qr$ **g** $p(q + r)$ **h** $2qr + 5pq$

i $\frac{q + r}{p}$ **j** $\frac{pq}{r}$ **k** $p + \frac{q}{r}$ **l** $\frac{2p}{3} + \frac{3q}{2} + \frac{5r}{2}$

4 Find the odd one out when $x = 8$

| $x + 6$ | $21 - x$ | $2x - 2$ |

5 Find the value of x that makes each of these expressions equal.

| $2x + 1$ | $22 - x$ | $3x - 6$ |

6 Find the values of x and y that make each of these expressions equal.

| $x + 2y + 16$ | $2x + 2y + 10$ | $x + 3y + 7$ |

Review

1 An apple costs a cents and a banana costs b cents.

Write an expression for the cost in cents of:

a 8 apples

b 5 bananas

c 8 apples and 5 bananas

d 15 apples and 9 bananas.

2 Simplify these expressions.

a $5h + 8h + 3h$ **b** $6x - 8x + 7x$ **c** $4b + 3c + 5b + 2c$

d $6f - 2g + 5f + g$ **e** $4 + 5x + 2y - 3x$ **f** $2xy - y + 5yx$

g $7x - 5y + 2x$ **h** $5ab + 2bc + 4ac - 2ba$

3 Copy and complete this algebraic wall.

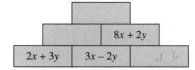

4 Write an expression for the perimeter of each shape.

a

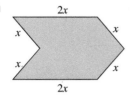

b

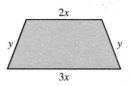

5 Multiply out the brackets.

   a $6(x + 3)$ **b** $2(x - 7)$ **c** $5(3 - x)$ **d** $3(x + y + 5)$

   e $8(2x + 3)$ **f** $3(4x - 1)$ **g** $9(2 - 7x)$ **h** $5(4a - 3b + 6c)$

6 Expand the brackets and simplify.

   a $3(5x + 4) + 2(6x - 3)$ **b** $5(5 - 2x) + 3(7 - 3x)$

7 If $x = 8$ and $y = 6$ find the value of:

   a $2x + 3y$ **b** $20 - 2x$ **c** $x(y - 3)$ **d** $2xy$

   e $9(2x - y)$ **f** $\frac{x}{4} + y$ **g** $\frac{xy}{12}$ **h** $\frac{3x}{y} + \frac{4y}{x}$

8 Find the value of x that makes each of these expressions equal.

   $\boxed{4x + 1}$ $\boxed{2x + 13}$ $\boxed{5x - 5}$

Shapes and geometric reasoning 1

Angles in action

Shape and angles are used in many jobs such as architecture, engineering, design and construction.

The person in this picture is using an instrument known as a theodolite. This measures angles very precisely.

Knowing the precise angles allows surveyors to draw accurate maps and plans based on triangles.

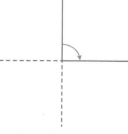

3.1 Angles

Angles are measured in degrees.

One full turn is 360 degrees (360°).

$\frac{1}{2}$ turn is 180°.

$\frac{1}{4}$ turn is 90°.

A **right angle**.

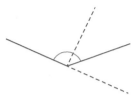

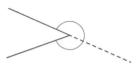

An **acute angle** is less than 90°.

An **obtuse angle** is between 90° and 180°.

A **reflex angle** is between 180° and 360°.

Worked example 1

Estimate the size of this angle.

Use straight line (180°), right angle (90°) and half right angle (45°) to help estimate the angle size.

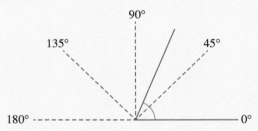

The angle is between 45° and 90°. It is about 70°.

Worked example 2

Estimate the size of this angle.

Use straight line (180°) and $\frac{3}{4}$ turn (270°) to help estimate the angle size.

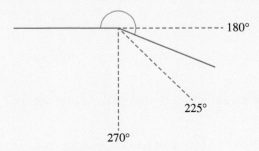

The angle is between 180° and 225°. It is about 200°.

Exercise 3.1

1 Here are eight angles.

Write down what type of angle each one is.

a

b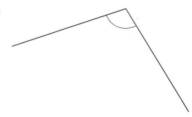

c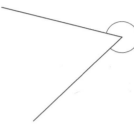

d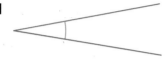

e

f

g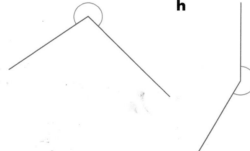

h

2 Estimate the size of each of the angles in question **1**.

 3 Sami draws a diagram to show how two acute angles fit together to make an obtuse angle.

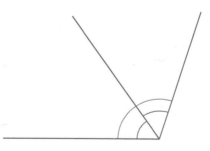

a Show how two acute angles could fit to make a different acute angle.

b Is it possible to make a reflex angle from two acute angles? Explain your answer.

c Show what other type of angle can be made from two acute angles.

> An acute angle is less than 90°. Can two of them add up to make a reflex angle greater than 180°?

3.2 Calculating angles

Angles at a point make a full turn. They add up to 360°.

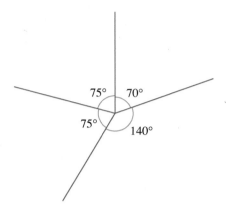

$$75° + 75° + 70° + 140° = 360°$$

Angles on a straight line add up to 180°.

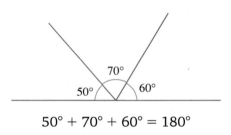

$$50° + 70° + 60° = 180°$$

Worked example 1

Calculate angle a.

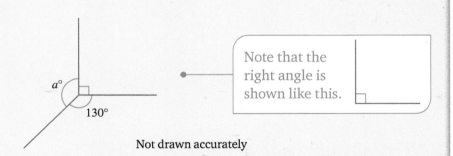

Note that the right angle is shown like this.

Not drawn accurately

$$90 + 130 = 220$$

$$360 - 220 = 140 \qquad \text{angles at a point add up to } 360°$$

$$a = 140$$

Worked example 2

The four angles shown are all equal. Find the size of angle *x*.

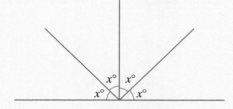

$180 \div 4 = 45$

$x = 45$ angles on a straight line add up to $180°$

Each angle is 45°.

Using letters to label shapes and angles

This is triangle *ABC*.

Each corner (or **vertex**) is labelled with a letter.

The side in red is side *AB*.

The marked angle is angle *ACB* or *BCA*.

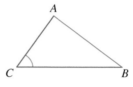

Worked example 3

In this diagram angle *DBC* = 65°

Find angle *ABD*.

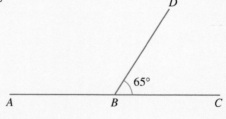

Not drawn accurately

$180 - 65 = 115$
Angle *ABD* = 115°

Vertically opposite angles

When two straight lines cross, the angles that face each other are called **vertically opposite** angles. This is because they are opposite each other at a **vertex**.

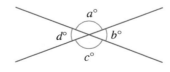

Angles *a* and *c* are vertically opposite.

Angles *b* and *d* are also vertically opposite.

Here is a proof that vertically opposite angles are equal.

In the diagram above, a and c are vertically opposite angles.

$a + d = 180$ \qquad angles on a straight line

Therefore $a = 180 - d$

$c + d = 180$ \qquad angles on a straight line

Therefore $c = 180 - d$

a and c are both equal to $180 - d$, so they must be equal to each other.

Therefore $a = c$

Similarly, you can show that angles b and d are equal.

Exercise 3.2

1 Calculate the size of each angle marked with a letter.

a
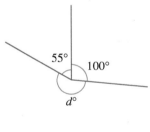

b
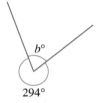

c

d
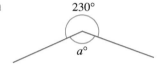

e
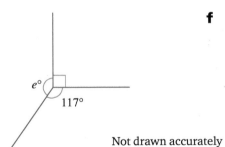

f
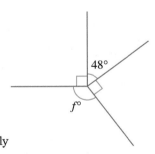

Not drawn accurately

2 Calculate the size of each angle marked with a letter.

a
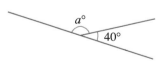

b
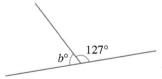

c
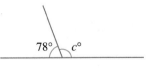

d
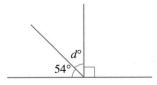

e
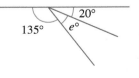

f
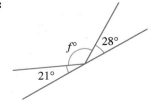

Not drawn accurately

3 Calculate the size of each angle marked with a letter.

a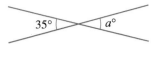

b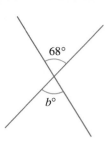

c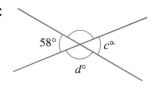

Not drawn accurately

4 *AB* and *CD* are straight lines. Angle *AMD* = 83°

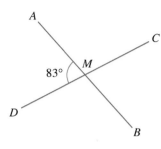

Not drawn accurately

    a Write down the size of angle *CMB*.

    b Calculate angle *BMD*.

5 Angles *PMQ*, *QMR* and *RMS* are all the same size. Angle *PMS* is twice as big as each of them. Work out the size of each angle.

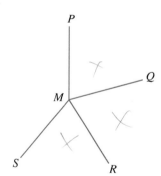

> Use letters and let angle *PMQ* = x.
> Write down the size of *QMR*, *RMS* and *PMS* in terms of x and make an equation which you can solve.

Not drawn accurately

6 The angles in each of these diagrams are equal. Find the size of each angle.

a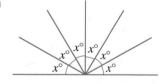

b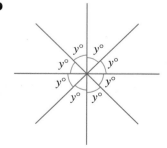

3.3 Triangles

A triangle has three sides and three angles.

The angles in a triangle add up to 180°.

$a° + b° + c° = 180°$

Worked example 1

Find angle x.

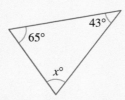

$65 + 43 + x = 180$ angles in a triangle add up to 180°

$180 - 65 - 43 = 72$

$x = 72$

Worked example 2

Find angle y.

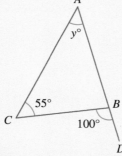

Angle $ABC = 80°$ angles on a straight line add up to 180°

$y = 180 - 55 - 80$ angles in a triangle add up to 180°

$y = 45$

Isosceles triangles

An isosceles triangle has two equal sides. It also has two equal angles.

If you know one angle in an isosceles triangle you can work out the other two.

Worked example 3

Find angles p and q.

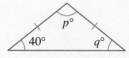

$q = 40$	in an isosceles triangle the two base angles are equal
$p = 180 - 40 - 40$	angles in a triangle add up to $180°$
$p = 100$	

Worked example 4

Find angle t.

$180 - 36 = 144$	therefore the two angles at the base add up to $144°$
$144 \div 2 = 72$	the two angles are equal
$t = 72$	

Other types of triangle

An **equilateral triangle** has three equal sides and three equal angles.

Each angle is $180 \div 3 = 60°$

A **right-angled triangle** has one angle equal to $90°$.

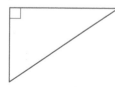

A **scalene triangle** is one with no equal sides and no equal angles.

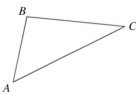

Worked example 5

This is a right-angled isosceles triangle. Find angle x.

$180 - 90 = 90$ therefore the two angles at the base add up to $90°$

$x = 90 \div 2$ the two angles are equal

$x = 45$

Exercise 3.3

1 Measure the sides of each of these triangles.

Choose the word from this list to describe each triangle.

Scalene   **Right-angled**   **Isosceles**   **Equilateral**

a
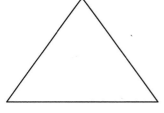

b
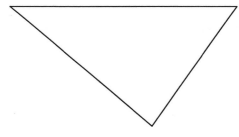

c
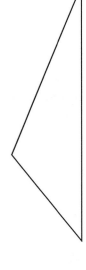

d
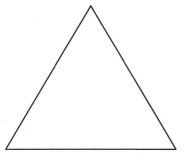

e
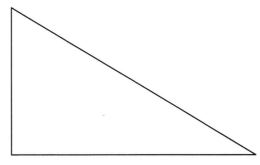

2 In these triangles find the angles marked with letters.

a
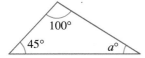

b
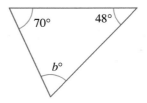

c
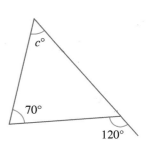

d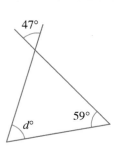

47°

59°

d°

e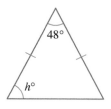

e°

125°   130°

f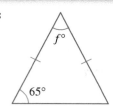

f°

65°

g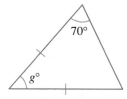

70°

g°

h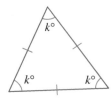

48°

h°

i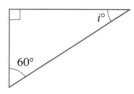

i°

60°

j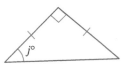

j°

k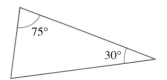

k°

k°   k°

Not drawn accurately

3 Show that this is an isosceles triangle. Explain the steps of your working.

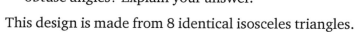

75°

30°

4 a Draw an isosceles triangle with no obtuse angles.

b Draw an isosceles triangle with one obtuse angle.

c Is it possible to draw an isosceles triangle with two obtuse angles? Explain your answer.

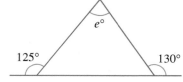
Each obtuse angle is greater than 90°. Can you have two obtuse angles in a triangle where the angles add up to 180°?

5 This design is made from 8 identical isosceles triangles.

a Find angle x.

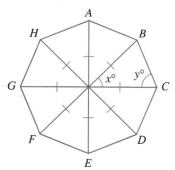

The angles at the centre are all equal and they add up to 360°.

b Find angle y.

3.4 Quadrilaterals

A **polygon** is a two-dimensional shape with straight sides. A **quadrilateral** is a polygon with four sides. Here are some examples of quadrilaterals.

Some shapes have sides that are **parallel**. Parallel lines continue in the same direction and would never meet. Matching arrows show sides that are parallel to each other.

There are some special types of quadrilateral that you need to know.

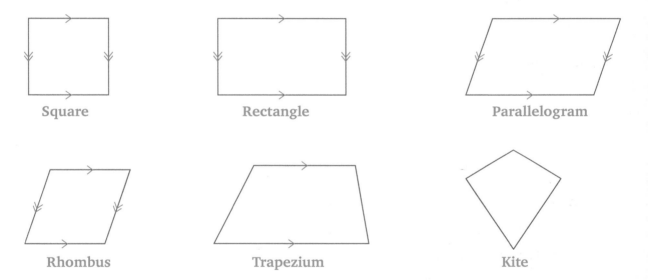

| Square | Rectangle | Parallelogram |

| Rhombus | Trapezium | Kite |

Angles in a quadrilateral

A line joining two **vertices** across a quadrilateral is called a **diagonal**.

A diagonal cuts any quadrilateral into two triangles.

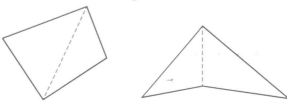

The angles in a triangle add up to 180°.

The angles in a quadrilateral add up to 180° × 2 = 360°

Worked example 1

Three of the angles in a quadrilateral are 48°, 67° and 128°.

Find the fourth angle.

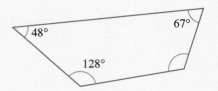

$48 + 67 + 128 = 243$ these three angles add up to 243°

$360 - 243 = 117$ angles in a quadrilateral add up to 360°

The fourth angle is 117°.

Worked example 2

ABCD is a quadrilateral.

The angles *ABC*, *DAB* and *BCD* are all equal to *x*.

Angle *CDE* = 48°.

Find *x*.

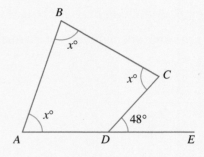

Angle $ADC = 180 - 48 = 132°$ angles on a straight line add up to 180°

$360 - 132 = 228°$ angles in a quadrilateral add up to 360°

$x = 228 \div 3 = 76$

Exercise 3.4

1 Find the missing angles in these quadrilaterals.

a

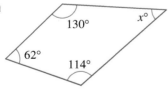

b

c

Not drawn accurately

2 Find angle *BCD*.

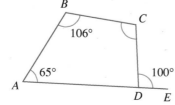

Not drawn accurately

3 Choose the correct word from the following list to describe each of these quadrilaterals.

 Kite   Parallelogram   Rectangle   Rhombus   Square   Trapezium

a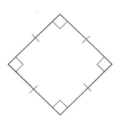

b

c

d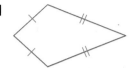

e

4 Work out the angles marked with letters.

a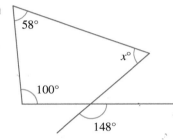
58°
$x°$
100°
148°

b
$y°$
116°
$y°$
Not drawn accurately

5 This diagram shows a quadrilateral with one obtuse angle.

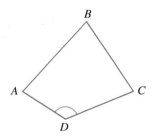
B

A   *C*

D

> You may need to try some sketches to find out what is possible.

Is it possible to draw the following quadrilaterals?
In each case draw a diagram or explain your answer carefully.

a A quadrilateral with two obtuse angles.

b A quadrilateral with three obtuse angles.

c A quadrilateral with four obtuse angles.

d A quadrilateral with no obtuse angles.

> Each obtuse angle is greater than 90° so what would four obtuse angles add up to?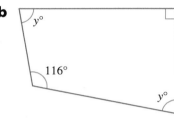

 6 Write down all the types of quadrilateral that fit these descriptions.

Part **a** is done for you as an example.

a A quadrilateral with four right-angled vertices.   **Answer:** Rectangle or square

b A quadrilateral with two pairs of equal sides.

> The two equal sides could be opposite each other or next to each other.

  c A quadrilateral with four equal sides.

  d A quadrilateral with two pairs of parallel sides.

  e A quadrilateral with four equal angles.

  f A quadrilateral in which the opposite angles are equal.

  g A quadrilateral with one pair of parallel sides.

  h A quadrilateral in which one pair of opposite angles are equal.

3.5 Symmetry

Symmetry occurs in many places in nature, art and buildings. Here are some real-life examples of symmetry.

You need to know about two types of symmetry – **line symmetry** and **rotation symmetry**.

Line symmetry

This arrow has a line of symmetry.

The line of symmetry is shown by the dotted line.

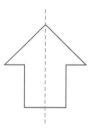

A line of symmetry is sometimes called a **mirror line**. If you place a mirror along the line the reflection looks the same as the original shape.

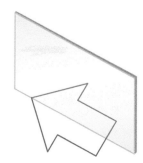

You can also check for symmetry by drawing the shape on paper and folding along the line of symmetry. The two halves should match exactly.

A rectangle has **two** lines of symmetry.

They are shown by the two dotted lines.

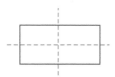

This shape has **no** lines of symmetry.

Worked example 1

How many lines of symmetry do these shapes have?

a

b

c

a This shape has **two** lines of symmetry.

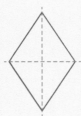

b This shape has **no** lines of symmetry.

c This shape has **four** lines of symmetry.

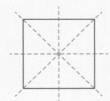

Rotation symmetry

This letter has rotation symmetry of **order 2**.

If you rotate the letter through one complete turn, it fits on to itself two times, so this is order 2.

Position 1   Position 2   Position 1

This shape has rotation symmetry of order 3.

If you rotate the shape through one complete turn, it fits on to itself three times, so this is order 3.

Worked example 2

What is the order of rotation symmetry of these shapes?

a 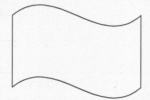   b    c

a This shape has rotation symmetry of order 2.

b This shape has rotation symmetry of order 1. (Sometimes this is described as having no rotation symmetry.)

c This shape has rotation symmetry of order 4.

Properties of triangles, quadrilaterals and other polygons

This table shows the special triangles and quadrilaterals you have met in this chapter so far.

The table also shows some other polygons.

Name	Shape	Sides	Angles	Line symmetry	Order of rotation symmetry
Scalene triangle		All sides different lengths	All angles different	No lines of symmetry	Rotation symmetry order 1
Isosceles triangle		Two equal sides	Two equal angles	One line of symmetry	Rotation symmetry order 1
Equilateral triangle		Three equal sides	All angles 60°	Three lines of symmetry	Rotation symmetry order 3
Square		Four equal sides	All angles 90°	Four lines of symmetry	Rotation symmetry order 4
Rectangle		Two pairs of equal sides	All angles 90°	Two lines of symmetry	Rotation symmetry order 2
Parallelogram		Two pairs of equal sides	Opposite angles equal	No lines of symmetry	Rotation symmetry order 2
Rhombus		Four equal sides	Opposite angles equal	Two lines of symmetry	Rotation symmetry order 2
Trapezium		All sides different lengths	All angles different	No lines of symmetry	Rotation symmetry order 1
Isosceles trapezium		Two equal sides	Two pairs of equal angles	One line of symmetry	Rotation symmetry order 1
Kite		Two pairs of equal sides	One pair of equal angles	One line of symmetry	Rotation symmetry order 1
Regular pentagon		Five equal sides	Five equal angles	Five lines of symmetry	Rotation symmetry order 5
Regular hexagon		Six equal sides	Six equal angles	Six lines of symmetry	Rotation symmetry order 6
Regular octagon		Eight equal sides	Eight equal angles	Eight lines of symmetry	Rotation symmetry order 8

Exercise 3.5

1 Copy each of these shapes and draw in any lines of symmetry.

a
b
c
d
e

2 For each of these shapes write down the order of rotation symmetry.

a N
b
c
d S
e

3 Here are some logos. Copy and complete the table showing their symmetry.

a
b
c
d

e
f
g
h

	Lines of symmetry	Order of rotation symmetry
a		
b		
c		
d		
e		
f		
g		
h		

4 Copy these grids on to squared paper.

Shade more squares to make the red lines into lines of symmetry.

a 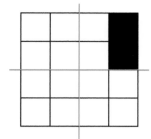   **b** 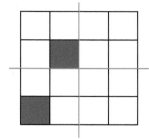   **c**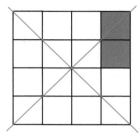

 5 Make three copies of this grid.

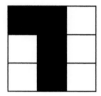

a On one copy shade one more square to give one line of symmetry.

> There is more than one way of doing this.

b On another copy shade one more square to give rotation symmetry of order 2 and no lines of symmetry.

c On the third copy shade one more square to give rotation symmetry of order 1 and no lines of symmetry.

 6 For each description try to draw a shape with the given symmetry.

If it is not possible, write 'Cannot be done'.

> Look at your answers to the earlier questions and then try drawing some sketches to see what is possible.

a 1 line of symmetry and rotation symmetry of order 1.

b 4 lines of symmetry and rotation symmetry of order 1.

c No lines of symmetry and rotation symmetry of order 4.

d 4 lines of symmetry and rotation symmetry of order 2.

e 2 lines of symmetry and rotation symmetry of order 4.

f 2 lines of symmetry and rotation symmetry of order 2.

Review

1 Write down what type of angles these are.

a    **b**    **c**

2 Estimate the size of these angles.

a

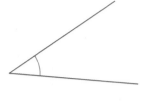

b

c

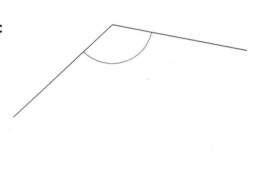

3 Work out the size of each angle marked with a letter.

a

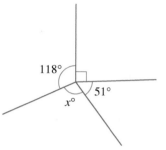

118°
51°
$x°$

b

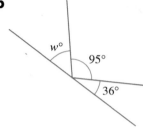

$w°$
95°
36°

c

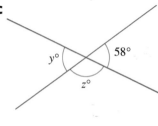

$y°$
58°
$z°$

d

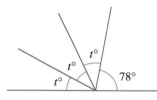

$t°$
$t°$
$t°$
78°

e

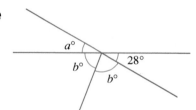

$a°$
$b°$
28°
$b°$

Not drawn accurately

4 Work out the missing angles in these shapes.

a

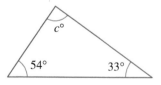

$c°$
54°
33°

b

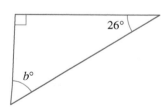

26°
$b°$

c

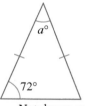

$a°$
72°

Not drawn accurately

5 *PQR* is a triangle.

Angle *RPQ* = 62°

Angle *PQR* = 51°

a Write down what type of triangle *PQR* is.

b Calculate angle *PRQ*.

c Calculate angle *PQS*.

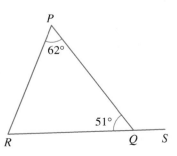

P
62°
51°
R
Q   S

Not drawn accurately

6 Work out the missing angles in these shapes.

a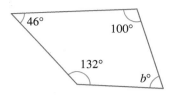

b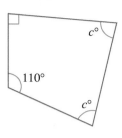

c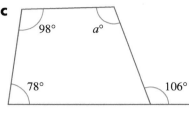

Not drawn accurately

7 Copy these shapes and draw in any lines of symmetry.

a

b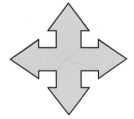

c

d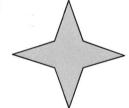

e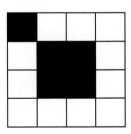

8 For each shape in question **7** write down the order of rotation symmetry.

9 Make three copies of this grid.

a On one copy shade three more squares to make a shape with rotation symmetry of order 4 and four lines of symmetry.

b On another copy shade three more squares to make a shape with rotation symmetry of order 2 and no lines of symmetry.

c On the third copy shade three more squares to make a shape with one line of symmetry.

4 Fractions

Learning outcomes

- Simplify fractions by cancelling common factors and identify equivalent fractions.
- Change an improper fraction to a mixed number.
- Know that in any division where the dividend is not a multiple of the divisor there will be a remainder.
- Add and subtract two simple fractions.
- Find fractions of quantities (whole number answers); multiply a fraction by an integer.
- Compare two fractions using diagrams.

Sharing pizzas

The word fraction is from the Latin 'fractus' meaning 'broken'. The history of fractions goes back to at least the ancient Egyptians.

Fractions are used to represent parts of a whole. They tell us how many parts of a certain size there are. For example, the fraction $\frac{3}{4}$ tells us that there are four equal parts and we have three of them.

The number on the bottom of the fraction is called the denominator. It tells us how many equal parts the whole has been divided into. The number on the top of the fraction is called the numerator. It tells us how many of the equal parts we have, or are interested in.

Fractions can be very useful when dividing something, such as a pizza or a cake, between numbers of people.

Dividing three pizzas equally between four people is not easy. If you think about dividing each pizza into quarters and then sharing these equally it makes it much easier to give each person a fair share.

4.1 Equal parts

This rectangle is divided into 2 equal parts.

One part is shaded.

$\frac{1}{2}$ of the rectangle is shaded and $\frac{1}{2}$ of the rectangle is not shaded.

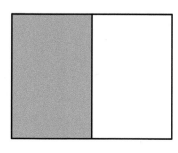

This rectangle is also divided into 2 parts.

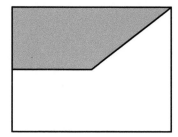

The two parts are not equal in size, so this time the fraction shaded is not $\frac{1}{2}$

You can add some lines to the rectangle to find out what fraction is shaded.

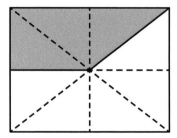

This now shows that there are 8 equal parts and 3 of them are shaded. So the fraction shaded is $\frac{3}{8}$

Before you decide what fraction is shaded, you should always make sure that you are working with equal parts.

Remember: The number on the bottom of the fraction is called the denominator. It tells you how many equal parts the whole has been divided into. The number on the top of the fraction is called the numerator. It tells you how many of the equal parts you have, or are interested in.

Worked example 1

What fraction of this circle is shaded?

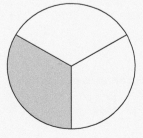

The circle is divided into 3 equal parts with 1 part shaded.

So $\frac{1}{3}$ (one-third) of the circle is shaded.

($\frac{2}{3}$ (two-thirds) of the circle is not shaded.)

Worked example 2

What fraction of this rectangle is shaded?

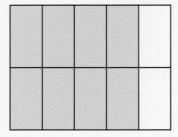

The rectangle is divided into 10 equal parts and 8 are shaded.

So $\frac{8}{10}$ of the rectangle is shaded.

There is another possible correct answer.

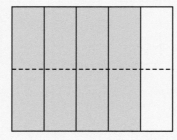

You can think of the rectangle as being divided into 5 equal parts, as shown. Four of these are shaded.

So $\frac{4}{5}$ of the rectangle is shaded.

When one shaded area can be described by two fractions in this way, the fractions are said to be equivalent to each other.

So, $\frac{8}{10}$ and $\frac{4}{5}$ are called equivalent fractions.

You will work with more equivalent fractions later.

Exercise 4.1

1 What fraction of each of these shapes is shaded?

a

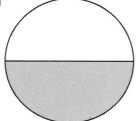

b

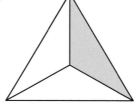

c

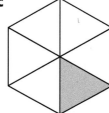

d 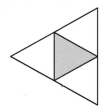 **e** 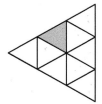 **f**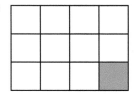

2 What fraction of each of these shapes is shaded?

a 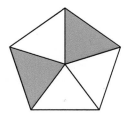 **b** 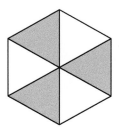 **c** 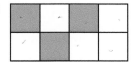 **d**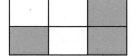

3 What fraction of each of these shapes is shaded?

a 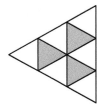 **b** 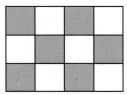 **c**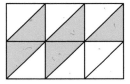

4 Copy this rectangle.

    a How many equal parts is the rectangle divided into?

    b Shade $\frac{5}{12}$ of the rectangle.

    c What fraction of the shape is not shaded?

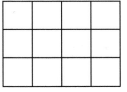

5 Copy this square.

    a How many equal parts is the square divided into?

    b Shade $\frac{7}{8}$ of the square.

    c What fraction of the square is not shaded?

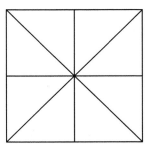

6 Kalil says that $\frac{1}{3}$ of this rectangle is shaded.

  Is he correct?

  Give a reason for your answer.

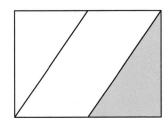

7 What fraction of each of these shapes is shaded?

a 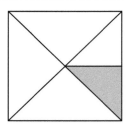 **b** 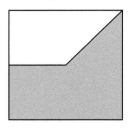 **c** 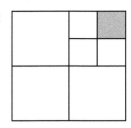 **d**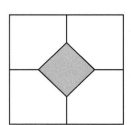

> Copy the diagrams. Add lines so that each diagram is divided into equal parts.

8 Make several copies of this square.

How many different ways can you join the dots to divide this square into two equal parts?

Use only straight lines and all lines must join dots.

> You may use more than one line on each diagram but they must join the dots. So, this is allowed

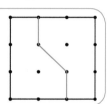

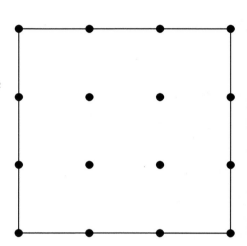

4.2 Equivalent fractions

You can find equivalent fractions by multiplying or dividing the numerator and denominator of a fraction by the same number.

Worked example 1

Find two fractions that are equivalent to $\frac{3}{4}$

$$\frac{3}{4} \overset{\times 2}{\underset{\times 2}{\frown}} \frac{6}{8} \qquad \frac{3}{4} \overset{\times 3}{\underset{\times 3}{\frown}} \frac{9}{12}$$

Remember that both the numerator and denominator must be multiplied by the same whole number.

Here they are multiplied by 2 and 3.

The equivalent fractions are $\frac{6}{8}$ and $\frac{9}{12}$

There are lots of other possible fractions that are equivalent to $\frac{3}{4}$

Worked example 2

Find two fractions that are equivalent to $\frac{8}{20}$

$$\frac{8}{20} \overset{\div 2}{\underset{\div 2}{\curvearrowright}} \frac{4}{10}$$

$$\frac{8}{20} \overset{\div 2}{\underset{\div 2}{\curvearrowright}} \frac{4}{10} \overset{\div 2}{\underset{\div 2}{\curvearrowright}} \frac{2}{5} \quad \text{dividing by 2 and then by 2 again has the same result as dividing by 4}$$

The equivalent fractions are $\frac{4}{10}$ and $\frac{2}{5}$

There are lots of other possible fractions that are equivalent to $\frac{8}{20}$

Simplifying

Dividing the numerator and denominator has the effect of making the figures in the fraction smaller. This is called simplifying.

A fraction that cannot be simplified further is in its simplest form or its lowest terms.

Worked example 3

Write the fraction $\frac{18}{24}$ in its simplest form.

18 and 24 are both multiples of 2 and 3 so you can divide by 2 or 3. Neither of these would give $\frac{18}{24}$ in its simplest form. You can also divide 18 and 24 by 6.

$$\frac{18}{24} \overset{\div 6}{\underset{\div 6}{\curvearrowright}} \frac{3}{4}$$

$\frac{18}{24}$ in its simplest form is $\frac{3}{4}$

Worked example 4

Copy and complete the following:

$$\frac{7}{8} = \frac{\square}{40}$$

The denominator 8 has been multiplied by 5 to get the answer 40.

To find equivalent fractions you multiply or divide the numerator and denominator by the same whole number.

$$\frac{7}{8} \overset{\times 5}{\underset{\times 5}{\curvearrowright}} \frac{35}{40}$$

The answer is $\frac{7}{8} = \frac{35}{40}$

Exercise 4.2

1 Write each of the following fractions in its simplest form.

a $\frac{9}{12}$ **b** $\frac{4}{10}$ **c** $\frac{5}{15}$ **d** $\frac{4}{18}$

e $\frac{6}{9}$ **f** $\frac{8}{10}$ **g** $\frac{10}{20}$ **h** $\frac{6}{15}$

i $\frac{21}{28}$ **j** $\frac{15}{20}$ **k** $\frac{70}{80}$ **l** $\frac{30}{80}$

2 Write each of the following fractions in its simplest form.

a $\frac{6}{18}$ **b** $\frac{15}{24}$ **c** $\frac{14}{42}$ **d** $\frac{9}{15}$ **e** $\frac{18}{27}$

3 Some of the following fractions can be simplified and some cannot.
Choose the fractions that can be simplified and write them in their simplest form.

a $\frac{14}{21}$ **b** $\frac{15}{35}$ **c** $\frac{17}{34}$ **d** $\frac{7}{18}$

e $\frac{9}{24}$ **f** $\frac{16}{48}$ **g** $\frac{9}{42}$ **h** $\frac{12}{17}$

i $\frac{12}{45}$ **j** $\frac{30}{48}$ **k** $\frac{18}{35}$ **l** $\frac{24}{60}$

4 Copy and complete the following.

a $\frac{1}{3} = \frac{\square}{6}$ **b** $\frac{3}{4} = \frac{15}{\square}$ **c** $\frac{7}{8} = \frac{\square}{24}$

d $\frac{3}{8} = \frac{\square}{32}$ **e** $\frac{5}{6} = \frac{\square}{30}$ **f** $\frac{7}{12} = \frac{\square}{60}$

5 Match the following into four pairs of equivalent fractions.

$\frac{5}{8}$ $\frac{4}{12}$ $\frac{9}{21}$ $\frac{15}{24}$ $\frac{5}{15}$ $\frac{15}{35}$ $\frac{12}{20}$ $\frac{9}{15}$

6 There are three sets of equivalent fractions mixed up below.
Sort the fractions into the three sets.

$\frac{20}{30}$ $\frac{15}{20}$ $\frac{16}{20}$ $\frac{24}{30}$ $\frac{18}{24}$ $\frac{16}{24}$ $\frac{24}{32}$ $\frac{24}{36}$ $\frac{28}{35}$

7 Baby Lily is playing with some bricks.
She has three red bricks, four yellow bricks and two green bricks.

   a What fraction of her bricks is green?

   b What fraction of her bricks is yellow?

   c What fraction of her bricks is red? Give your answer in its simplest form.

8 Jamie lists some fractions that are equivalent to $\frac{1}{3}$
Here is the start of his list.

$\frac{2}{6},$ $\frac{4}{12},$ $\frac{8}{24},$ $\frac{16}{48},$ $\frac{32}{96},$ ...

   a What is Jamie doing to work out the equivalent fractions?

   b Jamie says that he will eventually list all the fractions that
      are equivalent to $\frac{1}{3}$ if he keeps on going in this way.
      Show that Jamie is not correct.

> Find a fraction equivalent to $\frac{1}{3}$ that is between any pair in the list.

   c Suggest a way that could be used to find all the fractions that are equivalent to $\frac{1}{3}$

4.3 Mixed numbers

A mixed number consists of a whole number and a fraction.

For example $2\frac{3}{4}$ is a mixed number.

You can write $2\frac{3}{4}$ as a single fraction.

Its numerator would be greater than its denominator.

This is then called an improper fraction.

You can use diagrams to help change between mixed numbers and improper fractions.

Worked example 1

Write $2\frac{3}{4}$ as an improper fraction.

Using diagrams, $2\frac{3}{4}$ is represented by

$1 = \frac{4}{4}$, so $2\frac{3}{4} = 2 + \frac{3}{4} = \frac{8}{4} + \frac{3}{4} = \frac{11}{4}$

$2\frac{3}{4} = \frac{11}{4}$

You can do this without using diagrams by multiplying the number of wholes by the denominator of the fraction. Then add the numerator of the fraction.

Worked example 2

Write $3\frac{5}{6}$ as an improper fraction.

First work out the number of sixths in the whole number part.

There are $3 \times 6 = 18$ sixths in the whole number part (as $\frac{6}{6} = 1$, $\frac{18}{6} = 3$).

Add on the fraction. $\frac{18}{6} + \frac{5}{6} = \frac{23}{6}$

$3\frac{5}{6} = \frac{23}{6}$

Worked example 3

Write $\frac{9}{4}$ as a mixed number.

Using diagrams:

$\frac{9}{4} \quad = \quad \frac{4}{4} \quad + \quad \frac{4}{4} \quad + \quad \frac{1}{4}$

Remember that $\frac{4}{4} = 1$

$\frac{9}{4} = 2\frac{1}{4}$

$1 \quad + \quad 1 \quad + \quad \frac{1}{4} \quad = 2\frac{1}{4}$

Changing improper fractions to mixed numbers using division

Divide the numerator by the denominator and adding the remainder as a fraction of the denominator.

Worked example 4

Write $\frac{9}{4}$ as a mixed number.

To find how many wholes there are, divide 9 by 4 (there are 4 quarters in each whole).

$9 \div 4 = 2$ with remainder 1

This tells you there are 2 wholes.

Write the remainder as a fraction with the number you divide by in the denominator.

$\frac{9}{4} = 2\frac{1}{4}$

Exercise 4.3

1 Write each of these mixed numbers as an improper fraction.

   a $1\frac{1}{2}$ **b** $3\frac{1}{4}$ **c** $3\frac{2}{5}$ **d** $4\frac{7}{10}$ **e** $5\frac{3}{8}$ **f** $2\frac{3}{4}$

2 Write each of these improper fractions as a mixed number.

   a $\frac{7}{4}$ **b** $\frac{5}{2}$ **c** $\frac{8}{3}$ **d** $\frac{10}{7}$ **e** $\frac{9}{8}$ **f** $\frac{11}{4}$

4.4 Comparing, adding and subtracting fractions

Comparing fractions

You can use diagrams to compare two or more fractions.

Worked example 1

This rectangle is divided into twelfths.

Use the diagram to write the fractions $\frac{1}{3}, \frac{1}{4}$ and $\frac{1}{6}$ in ascending order (smallest first).

$\frac{1}{3} = \frac{4}{12}$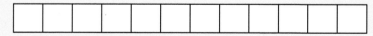

$\frac{1}{4} = \frac{3}{12}$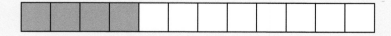

$\frac{1}{6} = \frac{2}{12}$

It is clear that the fractions are $\frac{1}{6}, \frac{1}{4}, \frac{1}{3}$ when written in ascending order of size.

Adding and subtracting fractions

Before you can add or subtract fractions they must be the same type of fraction. That is, they must have the same denominator.

The numerator tells you how many of each fraction you have. To add two fractions of the same type you just add the numerators. To subtract one fraction from another you subtract one numerator from the other.

Worked example 2

Work out $\frac{1}{3} + \frac{1}{6}$

Give your answer in its simplest form.

First make sure that both fractions to be added have the same denominator.

$\frac{1}{3}$ and $\frac{1}{6}$ do not have the same denominator. You need to use equivalent fractions that do have the same denominator.

$$\frac{1}{3} = \frac{2}{6}$$

$$\frac{1}{3} + \frac{1}{6} = \frac{2}{6} + \frac{1}{6} = \frac{3}{6}$$

This can be simplified.

$$\frac{3}{6} = \frac{1}{2}$$

So, $\frac{1}{3} + \frac{1}{6} = \frac{1}{2}$

Always show your working out. Do not jump straight from $\frac{2}{6} + \frac{1}{6}$ to $\frac{1}{2}$

Worked example 3

Work out $\frac{3}{4} - \frac{5}{12}$

First make sure that both fractions have the same denominator.

$\frac{3}{4}$ and $\frac{5}{12}$ do not have the same denominator. Use equivalent fractions that do have the same denominator.

$$\frac{3}{4} = \frac{9}{12}$$

$$\frac{3}{4} - \frac{5}{12} = \frac{9}{12} - \frac{5}{12} = \frac{4}{12}$$

This can be simplified.

$$\frac{4}{12} = \frac{1}{3}$$

So, $\frac{3}{4} - \frac{5}{12} = \frac{1}{3}$

Again, always show your working out. Never jump straight to the answer.

Exercise 4.4

1 Draw two copies of this diagram.

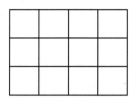

 a Shade in $\frac{3}{4}$ of your first diagram.

 b Shade in $\frac{7}{8}$ of your second diagram.

 c Which is larger, $\frac{3}{4}$ or $\frac{7}{8}$?

2 Draw three copies of this diagram.

 a Shade $\frac{5}{6}$ of your first diagram.

 b Shade $\frac{3}{4}$ of your second diagram.

 c Shade $\frac{2}{3}$ of your third diagram.

 d Use your diagrams to help you write the fractions $\frac{5}{6}$, $\frac{3}{4}$ and $\frac{2}{3}$ in order of size with the smallest first (ascending order).

3 This rectangle is divided into 12 equal squares.

 i Show each pair of fractions on your own copies of the rectangle above.

 ii Use your diagrams to decide which is the larger fraction of each pair.

 a $\frac{1}{6}, \frac{1}{4}$　　　　**b** $\frac{2}{3}, \frac{7}{12}$　　　　**c** $\frac{3}{4}, \frac{5}{6}$

4 Work these out. Write your answers in their simplest form.

 a $\frac{3}{8} + \frac{1}{8}$　　　　**b** $\frac{1}{8} + \frac{5}{8}$　　　　**c** $\frac{1}{4} + \frac{1}{4}$　　　　**d** $\frac{5}{6} + \frac{1}{6}$

5 Work these out. Write your answers in their simplest form.

 a $\frac{5}{6} - \frac{1}{6}$　　　　**b** $\frac{7}{8} - \frac{3}{8}$　　　　**c** $\frac{5}{8} - \frac{3}{8}$　　　　**d** $\frac{7}{12} - \frac{5}{12}$

6 Work these out. Write your answers in their simplest form.

 a $\frac{3}{4} + \frac{1}{8}$　　　　**b** $\frac{1}{3} + \frac{1}{6}$　　　　**c** $\frac{1}{4} + \frac{1}{12}$　　　　**d** $\frac{5}{8} + \frac{3}{16}$

 e $\frac{3}{10} + \frac{2}{5}$　　　　**f** $\frac{1}{3} + \frac{5}{12}$　　　　**g** $\frac{2}{7} + \frac{3}{14}$　　　　**h** $\frac{3}{5} + \frac{3}{10}$

 i $\frac{1}{2} + \frac{3}{8}$　　　　**j** $\frac{5}{8} + \frac{1}{4}$　　　　**k** $\frac{1}{4} + \frac{3}{20}$　　　　**l** $\frac{1}{2} + \frac{1}{3} + \frac{1}{12}$

7 Work these out. Write your answers in their simplest form.

 a $\frac{1}{2} - \frac{1}{4}$　　　　**b** $\frac{3}{4} - \frac{3}{8}$　　　　**c** $\frac{2}{5} - \frac{1}{10}$　　　　**d** $\frac{2}{3} - \frac{5}{12}$

 e $\frac{1}{4} - \frac{1}{16}$　　　　**f** $\frac{3}{7} - \frac{5}{14}$　　　　**g** $\frac{4}{5} - \frac{3}{10}$　　　　**h** $\frac{3}{5} - \frac{9}{20}$

 i $\frac{9}{10} - \frac{1}{2}$　　　　**j** $\frac{17}{20} - \frac{1}{4}$　　　　**k** $\frac{7}{12} - \frac{1}{4}$　　　　**l** $\frac{11}{24} - \frac{3}{8}$

8 Work these out. Give your answers as mixed numbers.

 a $\frac{3}{8} + \frac{7}{8}$　　　　**b** $\frac{5}{8} + \frac{3}{4}$　　　　**c** $\frac{4}{5} + \frac{9}{10}$

 d $\frac{5}{12} + \frac{2}{3}$　　　　**e** $\frac{7}{8} + \frac{3}{4}$　　　　**f** $\frac{5}{6} + \frac{7}{12}$

9 Use diagrams to put the following fractions in order of size with the smallest first (ascending order).

$\frac{5}{8}$, $\frac{2}{3}$, $\frac{7}{12}$

> Find the lowest common multiple of the denominators to decide how many equal parts you need in your diagrams.

10 Work out $3\frac{3}{5} + 2\frac{4}{5}$

> Add the whole number parts first.

4.5 Finding a fraction of an amount

Remember that fractions are all about dividing into equal parts.

The denominator tells you how many equal parts the whole has been divided up into.

So, to find $\frac{1}{4}$ of 64 you need to divide 64 up into 4 equal parts.

Work out $64 \div 4 = 16$

So, $\frac{1}{4}$ of $64 = 16$

'Of' in mathematics usually means 'multiplied by'.

$\frac{1}{4}$ of 64 is the same as $\frac{1}{4} \times 64 = \frac{1 \times 64}{4} = \frac{64}{4} = \frac{16}{1} = 16$

Worked example 1

Work out $4 \times \frac{2}{3}$

$4 \times \frac{2}{3} = \frac{4 \times 2}{3} = \frac{8}{3} = 2\frac{2}{3}$

$4 \times \frac{2}{3} = 2\frac{2}{3}$

Worked example 2

Work out one-third of 24.

One-third $= \frac{1}{3}$

$\frac{1}{3}$ of $24 = 24 \div 3 = 8$

Alternatively, $\frac{1}{3} \times 24 = \frac{24}{3} = 8$

Exercise 4.5

1 Find one-half of each of the following numbers.

    a 20   **b** 34   **c** 10   **d** 70   **e** 56   **f** 96

2 Find one-third of each of the following numbers.

    a 12   **b** 15   **c** 36   **d** 18   **e** 24   **f** 48

3 Find one-quarter of each of the following numbers

    a 12   **b** 16   **c** 24   **d** 36   **e** 84   **f** 72

4 Work out each of the following:

    a $\frac{1}{4}$ of 20   **b** $\frac{1}{3}$ of 21   **c** $\frac{1}{5}$ of 20   **d** $\frac{1}{6}$ of 24   **e** $\frac{1}{8}$ of 32   **f** $\frac{1}{9}$ of 45

5 Find the value of:

    a $\frac{1}{4}$ of 100 metres   **b** $\frac{1}{5}$ of \$100   **c** $\frac{1}{2}$ of 24 kg

6 Work out the following. Simplify your answer where possible.

    a $\frac{2}{7} \times 14$ **b** $3 \times \frac{2}{5}$ **c** $8 \times \frac{4}{5}$ **d** $5 \times \frac{5}{8}$ **e** $10 \times \frac{3}{4}$ **f** $6 \times \frac{5}{8}$

7 Work out each of the following:

    a $\frac{3}{8}$ of 40   **b** $\frac{2}{3}$ of 27   **c** $\frac{5}{6}$ of 36   **d** $\frac{2}{9}$ of 45   **e** $\frac{3}{4}$ of 48

8 Asha and Phil share a job. One week Asha works for 12 hours and Phil works for 8 hours.

    They are paid \$120 in total for the week's work.

    They share the pay fairly between them.

    a What fraction of the total hours does Asha work?

    b How much more pay does Asha receive than Phil?

> Start by working out the amount of pay Asha receives. Use your answer to part **a** to do this.

9 Mike has a tube of 12 sweets.

    He gives one-third of the sweets to a friend.

    He eats one-quarter of the rest.

    How many sweets does Mike have left?

> You can use a diagram showing 12 sweets. Shade $\frac{1}{3}$ of the sweets. Then shade $\frac{1}{4}$ of the rest.

10 Which is larger, $\frac{5}{6}$ of 24 or $\frac{3}{4}$ of 28? Give reasons for your answer.

Review

1 Johannes says that $\frac{1}{4}$ of this shape is coloured red.

    Give a reason why Johannes is not correct.

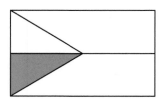

2 What fraction of each of these shapes is shaded?

    a 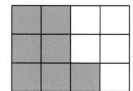   **b**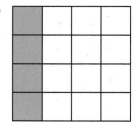

3 What fraction of this rectangle is shaded?

Give your answer in as many different ways as you can find.

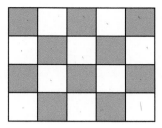

4 Copy and complete the following statements.

a $\frac{4}{9} = \frac{\square}{27}$ **b** $\frac{18}{48} = \frac{27}{\square}$

5 Write these fractions in their simplest form.

a $\frac{4}{12}$ **b** $\frac{6}{10}$ **c** $\frac{15}{24}$ **d** $\frac{16}{40}$ **e** $\frac{18}{27}$ **f** $\frac{21}{91}$

6 Which of these fractions does not have an equivalent in the list? Write the rest as pairs of equivalent fractions.

$\frac{4}{10}$, $\frac{3}{12}$, $\frac{6}{10}$, $\frac{6}{15}$, $\frac{4}{12}$, $\frac{18}{24}$, $\frac{15}{25}$, $\frac{3}{9}$, $\frac{7}{28}$

7 Write these improper fractions as mixed numbers.

a $\frac{24}{7}$ **b** $\frac{30}{8}$

8 Write these mixed numbers as improper fractions.

a $4\frac{1}{5}$ **b** $5\frac{7}{8}$

9 Work these out. Give your answers in their simplest form.

a $\frac{2}{3} + \frac{1}{6}$ **b** $\frac{5}{16} + \frac{1}{4}$ **c** $\frac{3}{4} - \frac{1}{8}$ **d** $\frac{7}{12} - \frac{1}{3}$

10 Work out:

a $\frac{3}{8} \times 32$ **b** $3 \times \frac{5}{6}$

5 Decimals

Learning outcomes

- Interpret decimal notation and place value.
- Round numbers to one decimal place, the nearest whole number or the nearest 10, 100 or 1000.
- Multiply and divide whole numbers and decimals by 10, 100 and 1000.
- Order decimals including measurements, changing these to the same units.
- Convert fractions to decimals to make comparisons.
- Convert terminating decimals to fractions.
- Recognise the equivalence of simple fractions and decimals.

Getting the point

Through history, people have used different ways to separate the whole number part from the fraction part of a number. Most cultures now use either a comma or a dot, but in the middle ages a bar was used above the fraction part.

In many sports, measurement of time needs to be as accurate as possible. In skiing, a very small amount of time can separate many competitors. In January 2013, in the women's Super-G event at St Anton am Arlberg, Austria, there was a gap of just 0.04 seconds between the winner, Tina Maze, and Anna Fenninger who finished second. The gap between third and fourth place was 0.01 seconds.

Times are measured to the nearest one-hundredth of a second using electronic sensors.

5.1 Place value

Place value and ordering

The decimal point separates the whole number or integer part from the fraction part.

The number 32.57 can be shown in a place value table like this.

Thousands	Hundreds	Tens	Units	.	Tenths	Hundredths	Thousandths
		3	2	.	5	7	

The value of the digit '3' is 30 (3 tens)
The value of the digit '2' is 2
The value of the digit '5' is 0.5
The value of the digit '7' is 0.07

Place value tables can be helpful when adding and subtracting decimals to get the digits in the right places.

Worked example

Write these numbers in ascending order.

3.45   2.6   3.01   2.58

> Remember that ascending order means in order of size with the smallest first.

You can write the numbers in a place value table to help.

Thousands	Hundreds	Tens	Units	.	Tenths	Hundredths	Thousandths
			3	.	4	5	
			2	.	6		
			3	.	0	1	
			2	.	5	8	

All the numbers begin with a digit in the units column in the table.

You compare these first.

There are two numbers starting with '2', so look at the tenths for these two numbers.

2.58 is smaller than 2.6 as '5' tenths is smaller than '6' tenths.

Do the same for the numbers starting with '3'.

3.01 is smaller than 3.45 as '0' tenths is smaller than '4' tenths.

This gives the order: 2.58   2.6   3.01   3.45

Exercise 5.1

1 Write these numbers in a place value table.

   a 23.5   **b** 37.67   **c** 2.09   **d** 321.12   **e** 400.20   **f** 103.07

2 What is the value of the underlined digit in each of these numbers?

   a 2.<u>3</u>47   **b** 1<u>5</u>.76   **c** <u>2</u>3.96   **d** 32.4<u>5</u>   **e** 42.0<u>2</u>   **f** 67.<u>8</u>2

3 Write each of these lists of numbers in ascending order.

   a 2.4   3.7   1.9   3.69   **b** 2.5   1.75   2.07   1.8

   c 7.06   7.5   6.7   6.05   **d** 3.2   3.14   5.001   2.17

   e 4.004   4.040   4.4   0.404   **f** 6.599   6.602   7.12   6.6

4 Write each of these lists of numbers in descending order.

> Remember that descending order means in order of size with the largest first.

   a 8.09   7.36   8.1   7.32   8.13

   b 10.26   8.99   9.85   11.76   11.24

   c 19.89   20.71   21.04   19.48   20.76

   d 15.55   15.555   15.505   15.055   15.5

   e 14.021   14.211   14.201   13.041   13.039

5 Here are the results of the 'World challenge – Zagreb' in the women's 200 metres race (September 2012).

Athlete	Time
Line KLOSTER	24.05
Charonda WILLIAMS	22.96
Allyson FELIX	22.35
Natalya NAZAROVA	23.94
Aleen BAILEY	22.95
Tatyana DEKTYAREVA	24.29
Anyika ONUORA	23.17
Aleksandra FEDORIVA	23.50

Start by listing the times in order of size. Remember that the winner has the smallest time.

The athlete with the lowest time was the winner.

List the athletes in their finishing positions.

6 Karim has these cards.

His teacher asks him to place all of them in the spaces on the classroom board.

He has to use them to make a number as close as possible to the number 4.

Show how Karim should arrange the cards.

7 Start in the shaded square.

Move in any direction (including diagonally) to a neighbouring square with the number closest to the number in the shaded square.

Now do the same for your new square, but you cannot visit a square twice.

The letters of the squares you visit will spell out a word.

What is the word?

P	M	L	E
0.51	0.02	0.71	0.53
N	A	D	G
0.62	0.74	0.8	0.53
O	R	N	P
0.69	0.71	0.49	0.58
U	N	L	I
0.68	0.6	0.31	0.5
D	E	D	O
0.77	0.81	0.55	0.02

Remember that the closest number may be above or below, and that you can only move one square at a time.

5.2 Rounding

Rounding to a given accuracy

Sometimes you do not need to know an exact value.

You can round figures to give an approximate value.

For example, the length of a book may be 29.76324. . . centimetres, but all we need to know is an approximate value for the length of the book.

You can round numbers to the nearest whole number, nearest 10, nearest 100, nearest 1000, and so on. Rounding to the nearest integer is the same as rounding to the nearest whole number.

You can also round numbers to a given number of decimal places.

In this chapter you will round to 1 decimal place. This means that there will be exactly one digit after the decimal point when rounded.

Worked example 1

Round 32.6 to:

a the nearest integer

b the nearest 10.

a 32.6 is between 32 and 33.

It is closer to 33 than to 32. So, 32.6 rounded to the nearest integer is 33.

b 32.6 is between 30 and 40.

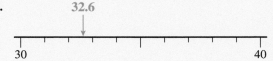

It is closer to 30 than to 40. So, 32.6 rounded to the nearest 10 is 30.

Worked example 2

Round 3.462 to 1 decimal place.

3.462 is between 3.4 and 3.5

It is closer to 3.5 than 3.4. So, 3.462 rounded to 1 decimal place is 3.5

Worked example 3

Round 6.5 to the nearest whole number.

6.5 is exactly halfway between 6 and 7 so there are two 'nearest' whole numbers.

When this happens 6.5 is rounded up to 7.

Similarly, 0.65 rounded to the nearest decimal place is 0.7

Sensible rounding

Sometimes you need to make decisions about rounding for yourself.

This will usually be because you cannot have a particular decimal as an answer.

Worked example 4

Kim buys some balls of wool to knit a jumper.

The pattern says she needs 4.2 balls of wool.

A ball of wool for Kim's jumper costs $3.20.

What is the cost of the wool for the jumper?

Kim needs to buy 5 balls of wool to have enough for the jumper.

Balls of wool are only sold in whole balls.

$5 \times \$3.20 = \$16$

Exercise 5.2

1 Round these to the nearest 10.

   a 78   **b** 26   **c** 343   **d** 475   **e** 198   **f** 1432.9

2 Round these to the nearest whole number.

   a 4.3   **b** 17.6   **c** 48.9   **d** 23.55   **e** 7.5   **f** 19.5

3 Round these to the nearest 100.

   a 349   **b** 252   **c** 1529   **d** 409   **e** 298   **f** 550

4 Round these to the nearest 1000.

   a 32 670   **b** 56 300   **c** 89 400   **d** 123 800   **e** 536 100   **f** 42 500

5 Round these to 1 decimal place.

    a 5.68   **b** 17.69   **c** 38.25   **d** 6.49   **e** 10.082   **f** 6.983

6 Round these distances to 1 decimal place.

    a 31.55 m   **b** 52.08 m   **c** 20.93 m   **d** 204.99 m   **e** 299.934 m   **f** 0.087 m

7 Work these out on your calculator.

Round your answers to the nearest whole number.

    a 7×3.2 **b** 8×13.2 **c** 6×12.8 **d** 6×4.7 **e** 3×3.2 **f** 5×3.3

8 Work these out on your calculator.

Round your answers to the nearest whole number.

    a $14.35 \div 5$ **b** $36.2 \div 4$ **c** $125.1 \div 8$ **d** $34.8 \div 3$ **e** $49.5 \div 5$ **f** $37.6 \div 8$

9 Bill says that 748 rounded to the nearest 10 is 75.

What mistake has Bill made?

10 Jane says 11.4 rounded to the nearest whole number is 12.

Is Jane correct?

Give a reason for your answer.

11 The table shows some of the tallest buildings in the world.

Building	Height (m)	
Burj Khalifa (formerly Burj Dubai), Dubai, The United Arab Emirates	828	
Taipei 101, Taipei, Taiwan	508	
World Financial Center, Shanghai, China	492	
Petronas Tower 1, Kuala Lumpur, Malaysia	452	
Petronas Tower 2, Kuala Lumpur, Malaysia	452	
Greenland Financial Center, Nanjing, China	450	
Sears Tower, Chicago, USA	442	
Guangzhou West Tower, Guangzhou, China	438	
Jin Mao Building, Shanghai, China	421	
Two International Finance Centre, Hong Kong	415	

(Source: http://www.infoplease.com)

Copy the table.

In the last column write the height of each building to the nearest 10 m.

12 Barak is making a chocolate cake.

His recipe states that 240 g of chocolate is needed.

Barak can buy 100 g bars that cost $1.30 each.

How much does the chocolate for the recipe cost?

13 Helen is carrying out a survey of students at her school.

The table shows the number of students in each year at her school.

She wants to survey $\frac{1}{10}$ of the people in each year at her school.

Copy the table.

In the last column write the number of students she should survey from each year.

Year group	Number of students in year	Number to be surveyed
7	120	
8	135	
9	124	
10	158	
11	143	

14 Haroon has a collection of stamps.

When the number of stamps is rounded to the nearest 10 it gives the same value as when the number of stamps is rounded to the nearest 100.

Give a possible value to the number of stamps that Haroon has.

15 John and his family are going on holiday.

They take three cases.

The total allowance for the cases is 45 kg.

John weighs the cases, and each case has a mass of 15 kg to the nearest kg.

John has to pay an extra charge at the airport as the cases exceed the total allowance of 45 kg.

Explain how this has happened.

> Think about the greatest mass that each case could have.

5.3 Multiplying and dividing by 10, 100 and 1000

Multiplying

When you multiply a number by 10 the units become tens, tenths become units and hundredths become tenths.

So every digit in the number moves one place to the left.

Tens	Units		Tenths	Hundredths
	8	.	5	6
8	5	.	6	

Multiply by 10 (arrows on each digit)

Multiplying by 100 is the same as multiplying by 10 and then by 10 again.

Multiplying by 1000 is the same as multiplying by 100 and then by 10, as $1000 = 100 \times 10$

Dividing

Dividing a number by 10 has the opposite effect to multiplying it by 10.

When you divide by 10, every digit in the number moves one place to the right.

Tens	Units	·	Tenths	Hundredths
8	5	·	6	
		·	8	5

Divide by 10 arrows showing 8→Units, 5→Tenths, 6→Hundredths giving 8 · 5 6

Dividing by 100 is the same as dividing by 10 and then by 10 again.

So every digit moves two places to the right.

Worked example 1

Multiply 8.56 by 100.

Hundreds	Tens	Units	·	Tenths	Hundredths
		8	·	5	6
8	5	6			

Multiply by 100 arrows

The answer is 856.

Worked example 2

Work out 5.2 × 1000

Thousands	Hundreds	Tens	Units	·	Tenths	Hundredths
			5	·	2	
5	2					
5	2	0	0			

Multiply by 1000 arrows

The answer is 5200.

You must remember to fill in any 'blank' columns to the left of the decimal point with '0's. When you do this you give the correct place value to the rest of the digits. In this example, the extra zeros make sure that the '2' stays in the hundreds column.

Worked example 3

Work out $5680 \div 100$

Thousands	Hundreds	Tens	Units	·	Tenths	Hundredths
5	6	8	0	·		
		5	6	·	8	0

The answer is 56.8

In the number 5680 the '0' gives 'place value' to the other digits.

Without the '0' the '8' would not be in the tens column. The '0' is not needed in the hundredths column in the final answer because it does not give place value to the other digits.

Exercise 5.3

You must not use a calculator in this exercise.

1 Work these out.

   a 2.3×10 **b** 0.7×10 **c** 1.8×100 **d** 4.35×100

   e 0.02×10 **f** 0.03×100 **g** 1.02×10 **h** 5.09×100

   i 20.1×10 **j** 2.5×1000 **k** 1000×0.05 **l** 1000×1.01

2 Copy each of the following.

Replace the ? with 10, 100 or 1000 to make the calculation correct.

   a $3.4 \times ? = 34$ **b** $6.02 \times ? = 602$ **c** $30.1 \times ? = 301$ **d** $0.85 \times ? = 85$

   e $2.01 \times ? = 2010$ **f** $0.12 \times ? = 120$ **g** $? \times 0.65 = 65$ **h** $? \times 1.03 = 10.3$

3 Work these out.

   a $23 \div 10$ **b** $0.7 \div 10$ **c** $180 \div 100$ **d** $435 \div 100$

   e $0.2 \div 10$ **f** $3020 \div 100$ **g** $1.2 \div 10$ **h** $5090 \div 100$

   i $20.1 \div 10$ **j** $2024 \div 1000$ **k** $32.8 \div 100$ **l** $7136 \div 1000$

4 Copy each of the following.

Replace the ? with 10, 100 or 1000 to make the calculation correct.

   a $3.4 \div ? = 0.34$ **b** $91 \div ? = 9.1$ **c** $0.3 \div ? = 0.03$ **d** $30 \div ? = 0.3$

   e $570 \div ? = 0.57$ **f** $523 \div ? = 5.23$ **g** $400 \div ? = 0.4$ **h** $50 \div ? = 0.05$

5 Copy each of the following. Replace the ? with $\times$ or $\div$ to make the calculation correct.

   a $5.2 ? 10 = 0.52$ **b** $0.85 ? 100 = 85$ **c** $451 ? 100 = 4.51$ **d** $1000 ? 0.02 = 20$

6 Use the numbers in the box to complete this calculation.

How many different ways can you find to do this?

? × 10 = ?

2.3		4		0.04
230	0.23		23	
	2300	0.4		40
				400

5.4 Calculating with decimals

Multiplying decimals

You can use facts you already know to help you multiply decimals.

Worked example 1

Work out 3 × 0.6 without using a calculator.

You know that 3 × 6 = 18

0.6 = 6 ÷ 10, so to make sure that both sides of the '=' sign stay equal you need to divide the right hand side by 10 also.

18 ÷ 10 = 1.8

So, 3 × 0.6 = 1.8

The diagram shows this.

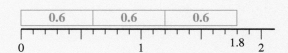

Worked example 2

Work out 8 × 0.12 without using a calculator.

You know 8 × 12 = 96

0.12 = 12 ÷ 100, so you need to divide 96 by 100.

96 ÷ 100 = 0.96

8 × 0.12 = 0.96

Worked example 3

Work out 12.3 × 7 without using a calculator.

Remember how to work this out.

```
  12.3
×    7
  86.1
   12
```

12.3 × 7 = 86.1

> Remember: there should be the same number of digits after the decimal point in the answer as there are in the calculation. Here there is one digit after the decimal point.

Dividing decimals

You can use facts you already know to help you divide decimals.

Adding and subtracting decimals

To add and subtract decimals you must make sure that the decimal points are lined up. You can do this by using a place value table. You can 'pair' numbers to help when you add decimals in your head.

Worked example 4

Work out $3.15 \div 5$ without using a calculator.

$$1.63$$
$$5\overline{)8.^31^15}$$

$3.15 \div 5 = 1.63$

> Remember: Dividing is the 'inverse operation' of multiplying. To check your answer you can multiply 1.63 by 5 to get 8.15

Worked example 5

Work out $5.2 \div 8$ without using a calculator.

> Add an extra '0' to 5.2 will help when you divide.

$$0.65$$
$$8\overline{)5.^52^40}$$

$5.2 \div 8 = 0.65$

Worked example 6

Work out **a** $3.5 + 5.8$ **b** $2.7 + 3.15$ **c** $14.4 - 3.7$ **d** $9.8 - 7.29$

a

Units	·	Tenths
3	·	5
+ 5₁	·	8
9	·	3

b

Units	·	Tenths	Hundredths
2	·	7	0
+ 3	·	1	5
5	·	8	5

> It is a good idea to put '0' in the space to make both numbers line up on the right.

c

Tens	Units	·	Tenths
1	³4̶	·	¹4
–	3	·	7
1	0	·	7

d

Units	·	Tenths	Hundredths
9	·	⁷8̶	¹0
– 7	·	2	9
2	·	5	1

> It is a good idea to put '0' in the space to make both numbers line up on the right.

Worked example 7

Work out $7.1 + 3.4 + 2.9 + 4.6$ without using a calculator.

You know that $0.1 + 0.9 = 1$, and that $0.4 + 0.6 = 1$

You can use this to 'pair' the numbers together.

$$7.1 + 3.4 + 2.9 + 4.6 \quad = \underline{7.1 + 2.9} + \underline{3.4 + 4.6}$$
$$= \quad 10 \quad + \quad 8$$
$$= \quad\quad 18$$

So, $7.1 + 3.4 + 2.9 + 4.6 = 18$

Exercise 5.4

You must not use a calculator in this exercise.

1 Work these out.

  a 3×0.4 **b** 5×0.7 **c** 0.6×8 **d** 0.8×6

  e 12×0.6 **f** 7×0.9 **g** 0.4×6 **h** 1.2×11

> Always think about your answer. Remember, if you multiply by a number greater than 1 the answer will be greater than your starting number. If you multiply by a number between 0 and 1 then your answer will be smaller.

2 Work these out.

  a 8.7×4 **b** 3.4×8 **c** 12.6×8 **d** 7.65×7 **e** 3.42×6 **f** 7.95×4

3 Work these out.

  a $7.65 \div 5$ **b** $9.84 \div 6$ **c** $4.35 \div 5$ **d** $3.6 \div 4$ **e** $2.7 \div 5$ **f** $4.4 \div 8$

4 Work these out.

  a $3.4 + 5.2$ **b** $13.7 + 8.1$ **c** $16.3 - 8.1$ **d** $15.3 - 4.2$

5 Work these out.

  a $6.3 + 4.9$ **b** $13.6 + 2.7$ **c** $15.7 + 7.4$ **d** $13.6 + 5.4$

6 Work these out.

  a $3.26 + 4.5$ **b** $7.1 + 4.39$ **c** $15.73 + 9.18$

  d $23.8 + 27.36$ **e** $2.356 + 4.725$ **f** $8.095 + 4.92$

7 Rafad says that $5.12 + 3.6 = 8.18$

  James says that $5.12 + 3.6 = 8.72$

  Who is correct? Give a reason for your answer.

8 Work these out.

  a $7.8 - 4.2$ **b** $9.6 - 6.3$ **c** $14.2 - 3.1$ **d** $8.7 - 6.7$

9   Work these out.

  a $17.7 - 5.9$ **b** $26.5 - 3.8$ **c** $14.6 - 4.8$ **d** $43.2 - 5.8$

10 Work these out.

  a $13.94 - 2.7$ **b** $16.9 - 4.35$ **c** $38.4 - 23.62$

  d $23.45 - 4.6$ **e** $23.685 - 12.745$ **f** $24.55 - 17.305$

11 Nadia says that $7.4 - 4.31 = 3.09$

  Sara says that $7.4 - 4.31 = 3.11$

  Who made the mistake? Explain the mistake she made.

12 Work these out.

  a $12.3 + 4.7$ **b** $3.7 + 4 + 2.3$ **c** $3.4 + 2.6 + 4.1 + 2.9$

  d $2.3 + 3.5 + 7.7 + 6.5$ **e** $4.2 + 4.6 + 3.8 + 5.4$ **f** $7.6 + 3.2 + 2.4 + 5.8$

13 Work out the total amounts for each of the following.

  a Six tins of paint, each containing 0.6 litres of paint.

  b Four jars, each containing 0.4 litres.

  c Seven packs, each with a mass of 1.2 kg.

  d Eight bottles, each containing 0.6 litres.

14 Cheese costs $12 per kg.

  How much does 0.3 kg cost?

15 Work these out.

  a 20×0.03 **b** 0.9×200 **c** 0.6×0.3 **d** 20×0.08
  > Remember that 20×0.03 $= 2 \times 10 \times 0.03$

16 Copy each of the following. Each ? represents a missing digit.

  Replace the ? with the digit that makes the calculation correct.

  a $?.6 + 4.? = ?2.8$ **b** $9.7 + ?.6 = ?3.3$ **c** $?3.?6 + 1?.3? = 25.78$
  d $8.? - ?.9 = 4.7$ **e** $1?.6 - 6.? = 7.8$ **f** $8?.04 - 48.? = ?0.64$
  > Take extra care when you have to carry a digit.

5.5 Changing between fractions and decimals

Converting fractions to decimals

To convert a fraction to a decimal you divide the numerator (top number) by the denominator (bottom number) on your calculator.

$\frac{4}{5}$ means $4 \div 5$ so $\frac{4}{5}$ written as a decimal is 0.8

Worked example 1

Convert the following fractions to decimals.

a $\frac{7}{8}$ **b** $\frac{9}{20}$

a $\frac{7}{8}$ means $7 \div 8$, so $\frac{7}{8}$ written as a decimal is 0.875

b $\frac{9}{20}$ means $9 \div 20$, so $\frac{9}{20}$ written as a decimal is 0.45

Worked example 2

Which is larger, $\frac{3}{5}$ or $\frac{13}{20}$?

Convert each of the fractions to a decimal.

$$\frac{3}{5} = 0.6 \qquad\qquad \frac{13}{20} = 0.65$$

0.65 is larger than 0.6, so $\frac{13}{20}$ is larger.

Converting decimals to fractions

To convert a decimal to a fraction, put the decimal into a place value table.

The digits from the decimal give you the numerator of the fraction.

The heading of the number column of the last digit gives you the denominator. For example, if the last digit is in the tenths column, the denominator is 10.

Simplify the resulting fraction if possible.

Worked example 3

Write the following decimals as fractions.

a 0.76 **b** 0.125

a Write the digits in a place value table.

Units	·	Tenths	Hundredths
0	·	7	6

The last digit is in the hundredths column, so put 100 in the denominator, and the digits 76 in the numerator.

Then simplify.

$$\frac{76}{100} = \frac{19}{25}$$

b Write the digits in a place value table.

Units	·	Tenths	Hundredths	Thousandths
0	·	1	2	5

The last digit is in the thousandths column, so put 1000 in the denominator, and the digits 125 in the numerator.

Then simplify.

$\frac{125}{1000} = \frac{1}{8}$

> You can check your answer by working backwards. $1 \div 8 = 0.125$

Exercise 5.5

1 Change these fractions to decimals.

a $\frac{3}{8}$ **b** $\frac{7}{8}$ **c** $\frac{3}{16}$ **d** $\frac{4}{25}$ **e** $\frac{9}{20}$ **f** $\frac{9}{10}$

g $\frac{9}{25}$ **h** $\frac{11}{20}$ **i** $\frac{1}{4}$ **j** $\frac{3}{4}$ **k** $\frac{4}{5}$ **l** $\frac{19}{20}$

2 Which of these fractions is closest to 0.57?

$\frac{11}{20}$ $\frac{3}{5}$ $\frac{29}{50}$

3 Use decimals to decide which fraction is larger from each of the following pairs.

a $\frac{3}{8}, \frac{7}{20}$ **b** $\frac{3}{5}, \frac{5}{8}$ **c** $\frac{7}{25}, \frac{3}{10}$

4 Write these decimals as fractions.

a 0.37   **b** 0.45   **c** 0.2   **d** 0.3   **e** 0.35   **f** 0.85

g 0.75   **h** 0.8   **i** 0.15   **j** 0.28   **k** 0.07   **l** 0.12

5 In 2011, Peter Glazebrook set a new world record for the heaviest onion.
It had a mass of 8.15 kg. Which of the following is equivalent to 8.15?

$8\frac{1}{15}$ $8\frac{1}{4}$ $8\frac{3}{20}$ $8\frac{1}{5}$

6 Write these numbers in order of size, smallest first.

a $\frac{3}{4}$ 0.8   0.69 $\frac{31}{50}$

b $\frac{1}{5}$ 0.15 $\frac{1}{4}$ 0.09

7 Pablo says 0.35 is larger than 0.4 because 35 is larger than 4.

Is he correct? Give a reason for your answer.

Review

1 What is the value of the underlined digit in each of the following numbers?

    a 2<u>3</u>7.67   **b** 52<u>9</u>.48   **c** 0.1<u>6</u>4   **d** 10<u>5</u>4.6

2 Write each of these lists in ascending order.

    a 4.3   4.03   4.65   4.16   4.079

    b 12.1   12.89   12.08   12.45   12.09

    c 0.17   0.71   0.071   0.171   0.177

3 The number of people attending a football match is 23 645.

    Round this number to the nearest

    a 1000   **b** 100   **c** 10

4 Round each number to 1 decimal place.

    a 4.09   **b** 5.65   **c** 752.05   **d** 99.04

5 Work these out without using a calculator.

    a 3.4×2 **b** 1.6×3 **c** 5×0.8 **d** 9×0.6

6 Evie wants to make 4 cushion covers. Each cushion cover needs 1.2 metres of material.

    How much material does she need altogether?

7 Work these out without using a calculator.

    a 5.2×10 **b** 6.04×100 **c** $432 \div 100$ **d** $5.2 \div 10$

8 Work these out without using a calculator.

    a $2.3 + 5.4$ **b** $1.8 + 2.5$ **c** $0.73 + 0.4$ **d** $23.56 + 9.8$

    e $9.8 - 3.5$ **f** $9.7 - 3.8$ **g** $16.46 - 9.7$ **h** $34.8 - 7.09$

9 Lily has a mass of 18.6 kg. Laura has a mass of 29.45 kg.

    a Find the difference in their masses.

    b What is the total mass of Lily and Laura?

10 Write these decimals as fractions.

    a 0.56   **b** 0.75   **c** 0.35

11 Write this list of numbers in ascending order of size.

    $\frac{4}{5}$ 0.74 $\frac{33}{50}$ 0.61

12 Peter's truck can carry a maximum load of 550 kg including passengers and driver.

    He has 5 cases each with a mass of 30 kg. He has 10 boxes each with a mass of 32 kg.

    Peter has a mass of 90 kg. Can the truck carry Peter, all 5 cases and all 10 boxes?

    Show all your working.

6 Processing, interpreting and discussing data

Chapter 18 covers collecting data
Chapter 10 covers presenting data

Learning outcomes

- Find the mode (or modal class for grouped data), median and range.
- Calculate the mean, including from a frequency table.
- Draw conclusions based on simple statistics.
- Compare two simple distributions using the range and the mode, median or mean.

Taller or shorter?

Which of the following groups of people is the taller?

It is almost impossible to make a comparison like this. There are lots of people in both groups. You can try pairing people of the same size in each group or look at the tallest and the least tall. But there may be more people in one group than the other, or there may be one particularly tall person in one of the groups.

It is much easier to use just one value to represent the heights for each group. This can be an average value. The mode, median and mean are examples of average values.

An average value alone is not usually enough to compare groups. It also helps to know what is happening around the average value. We use a measure of spread to do this. The range is an example of a measure of spread.

6.1 Mode and median

Finding the mode

The easiest average to find is the mode.

It is the value that occurs most frequently (most often).

Sometimes there is no mode because there is no number that occurs more frequently than the other numbers. Sometimes there may be more than one mode.

Worked example 1

Carlos asks his friends how many CDs they bought last week.

They reply:

0, 4, 3, 2, 5, 2, 6, 3, 2, 4, 5, 6, 7, 1, 2, 3, 1, 2

Find the modal number of CDs bought by Carlos's friends last week.

The 'modal number' of CDs means the mode of the number of CDs.

It usually helps to sort the numbers in order of size.

0, 1, 1, 2, 2, 2, 2, 2, 3, 3, 3, 4, 4, 5, 5, 6, 6, 7

It is now easy to see that there are more 2's than any other number.

So, more of Carlos's friends bought 2 CDs than any other number.

The modal number of CDs bought is 2.

Worked example 2

Find the mode of each of the following.

a 1, 3, 5, 6, 7, 9, 10, 12, 13

b 3, 4, 4, 5, 7, 7, 8, 10, 11

a As no value occurs more frequently than the others there is no mode.

b There are two modes, 4 and 7.

Finding the median

The median is the middle value of a set of data when all the values are listed in order of size.

Worked example 3

Find the median of:

0, 4, 3, 2, 5, 2, 6, 3, 2

First write the values in order of size:

0, 2, 2, 2, 3, 3, 4, 5, 6

There are 9 values. The one in the middle is the 5th one, which is shown in red.

The median is 3.

Worked example 4

Find the median of:

2, 5, 8, 9, 10, 11

These are already in order of size.

There are 6 values, so there is no middle value.

2, 5, 8, 9, 10, 11

Instead there are two values in the middle, which make a middle pair. These are shown in red.

The median is halfway between these values.

$$\frac{8 + 9}{2} = 8.5$$

So, the median is 8.5

Worked example 5

Find the median number of CDs bought by Carlos's friends in Worked example 1.

0, 1, 1, 2, 2, 2, 2, 2, 3, 3, 3, 4, 4, 5, 5, 6, 6, 7

There are 18 values. Eighteen is an even number so there is a middle pair.

The median is halfway between these values, but both values are 3.

So, the median is 3.

Exercise 6.1

1 Write down the mode and median for each of the following.

  a 2, 4, 5, 5, 6, 7, 8, 9, 9, 10, 10, 11, 11, 11, 12, 13, 14

  b 22, 25, 26, 26, 26, 27, 28, 30

  c 32, 32, 33, 34, 35, 35, 36, 37

  d 50, 51, 52, 53, 55, 56

2 Five houses were for sale at $89 000, $92 000, $135 000, $123 000 and $90 000.

  What is the median price?

3 A group of students got the following marks in a test:

  18, 15, 15, 17, 18, 13, 17, 16, 15, 19, 15, 18

  a What was the modal mark?

  b What was the median mark?

4 The ages, in months, of young children in a nursery are:

15, 18, 20, 16, 17, 20, 24, 15, 17, 16, 18

19, 14, 13, 22, 17, 16, 17, 22, 21, 17, 15

a Find the median age of the children.

b What is the modal age?

5 The prices of some shirts in a shop are:

$7.80, $8.50, $7.99, $10.20, $8.99, $6.99, $9.50, $8.20, $6.99

a What is the modal price?

b What is the median price?

c Which of the mode or median would be best to represent the prices of the shirts? Give a reason for your answer.

6 Here are the masses, in kg, of a group of baby boys and baby girls.

Boys	3.0	2.4	1.8	2.2	1.9	3.2	2.8	2.7	2.2	2.1	1.9
Girls	2.7	3.1	3.0	1.4	2.6	2.8	2.3	2.5	2.6	2.1	

Which group has the greater median mass?

7 Christina thinks of 4 numbers.

The smallest number is 5. The largest number is 9. The mode is 7.

Write down the 4 numbers.

8 Yasmin thinks of a set of 5 numbers.

The median is 2 lower than the mode.

List a possible set of 5 numbers.

> There is a mode. The median is two lower than the mode. This tells you that the two largest values are the same.

9 David thinks of a set of 5 numbers.

The smallest number is 4. The largest number is 10. The median is 7.

Write down a possible set of numbers.

> List the numbers you are given in order. Fill in the gaps so that 7 is the median.

6.2 Mean and range

Finding the mean

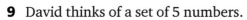

You can find the mean by adding all the values together and dividing the total by the number of values.

The mode, mean and the median are all measures of 'average', but when the term 'average' is used it usually refers to the mean value.

Worked example 1

Find the mean of these lengths.

4 m, 8 m, 5 m, 9 m, 3 m

$$\text{Mean} = \frac{4 + 8 + 5 + 9 + 3}{5} = \frac{29}{5} = 5.8\,\text{m}$$

Worked example 2

The mean mass of a team of 5 people is 72 kg.

a What is the total mass of the team?

b A player with a mass of 68 kg is replaced by a player with a mass of 73 kg.

What is the mean mass of the team now?

a Mean weight = $\dfrac{\text{total mass}}{\text{number of players}} = \dfrac{\text{total mass}}{5} = 72$

So, total mass = $5 \times 72 = 360$ kg

b New total mass = $360 - 68 + 73 = 365$ kg

New mean mass = $\dfrac{365}{5} = 73$ kg

Finding the range

The range is a measure of spread.

The range is the difference between the largest and smallest values in a set of data.

Worked example 3

Find the range of these lengths.

4 m, 8 m, 5 m, 9 m, 3 m

The range = 9 m − 3 m = 6 m

Exercise 6.2

1 Find the mean of the following:

    a 2, 6, 4, 9, 3, 4, 2, 1, 8, 3   **b** 12, 13, 16, 19, 20, 8, 5, 18, 14, 16

    c 4, 3, 1, 5, 0, 6, 10, 18, 6, 9, 10, 12   **d** 25, 30, 24, 21, 31, 39, 44, 26

2 Find the range for each set of numbers in question **1**.

3 The numbers of goals scored by a football club in 15 games were:

    0, 4, 2, 3, 2, 5, 1, 2, 1, 0, 3, 4, 2, 3, 1

    a Find the range of the number of goals.

    b Find the mean number of goals per game.

4 These are the number of tomatoes Asif gets from his plants.

    23, 45, 31, 52, 18, 16, 28, 32, 8, 24

    a What is the mean number of tomatoes per plant?

    b What is the range?

5 Here is a list of numbers:

2, 5, 6, 8, 9

a Find the mean of the numbers.

b Write down the range of the numbers.

10 is added to each of the numbers in the list.

c Find the new mean.

d Write down the range of the new numbers.

6 These are the ages, in years, of the players in a football team.

23, 27, 25, 31, 27, 28, 32, 21, 22, 24, 37

a Find the mean age of the players.

b What is the range?

c The reserve player is 39 years old. Work out the mean age for all 12 players.

7 The mean mass of the 11 players in a football team is 82 kg.

a What is the total mass of the team?

b A player with a mass of 86 kg is replaced with a player with a mass of 75 kg.

Find the new mean mass of the team.

> There are 4 numbers. They have a mode. The mean is less than the mode. This tells you that there are two equal values and the other two numbers are lower. Remember that you add the 4 numbers and divide the total by 4 to find the mean.

8 Adina thinks of a set of 4 whole numbers less than 10.

The mean of the numbers is 1 less than the mode.

Find a possible set of values for the 4 numbers.

9 The mean mass of the 15 players in a rugby team was 93 kg.

Daniel, who had a mass of 96 kg, was replaced by Thomas.

Afterwards the mean mass was 92 kg.

> Work out the total mass of the team with David and then the total mass of the team with Thomas.

What was Thomas's mass?

10 There are 9 shirts on a rack in a shop.

One of the shirts is sold for $28.

> Start by working out the total cost of the remaining 8 shirts.

The mean price of the remaining shirts is $19.

What was the mean price before the shirt was sold?

6.3 Using frequency tables

Reading frequency tables

It can be difficult to read a long list of values. A better method is to put the values into a table.

This type of table is called a **frequency table**.

Worked example 1

The table shows the number of books read by a class of students last month.

a How many students read 5 books?

b What was the modal number of books read?

c How many students are in the class altogether?

d What is the total number of books read by the class last month?

Number of books read	Frequency
0	3
1	5
2	5
3	7
4	5
5	2

a Two students read 5 books. This is shown in the final row of the table. Remember, the frequency tells you the number of students.

b The modal number of books read was 3, because 3 books had the greatest frequency.

c To find the total number of students in the class you add up the frequencies.

$$3 + 5 + 5 + 7 + 5 + 2 = 27$$

There are 27 students in the class.

d 3 students read 0 books = $3 \times 0 = 0$ books

5 students read 1 books = $5 \times 1 = 5$ books

5 students read 2 books = $5 \times 2 = 10$ books

7 students read 3 books = $7 \times 3 = 21$ books

5 students read 4 books = $5 \times 4 = 20$ books

2 students read 5 books = $2 \times 5 = 10$ books

Total number of books = $0 + 5 + 10 + 21 + 20 + 10 = 66$

This calculation can be done using the table and adding another column for the products.

Number of books read	Frequency	Product
0	3	0
1	5	5
2	5	10
3	7	21
4	5	20
5	2	10
Total number of books read =		66

Finding the mean from a frequency table

In section 6.2 you learned that the mean is found by adding all the values together and dividing the total by the number of values. This can be done using a frequency table.

Worked example 2

The table in Worked example 1 part **d** shows the total number of books read by a class of students last month.

Use the table to calculate the mean number of books read by the students last month.

The mean number of books read $= \dfrac{\text{total number of books read}}{\text{number of students}}$

$= \dfrac{66}{27}$

$= 2.4$

> This is the $\dfrac{\text{sum of the product column}}{\text{sum of the frequency column}}$

Worked example 3

A teacher records the number of children in the families of her students.

The results are shown in the table.

Number of children	Frequency
0	0
1	4
2	8
3	10
4	6
5	2

Find the mean number of children per family.

First, add an extra column to the right-hand side, and an extra row at the bottom of the table.

Number of children	Frequency	Product	
0	0	0	0 × 0
1	4	4	1 × 4
2	8	16	2 × 8
3	10	30	3 × 10
4	6	24	4 × 6
5	2	10	5 × 2
Totals	30	84	

The mean number of children per family $= \frac{84}{30} = 2.8$

Exercise 6.3

1 Chen records the number of people in cars as they drive past his school.

His results are shown in the frequency table.

Show your working when you answer these questions.

Number of people in car	Frequency
1	7
2	10
3	6
4	5
5	2

   a How many cars are there altogether?

   b How many cars had 2 people in?

   c What is the total number of people in cars with 2 people?

   d What is the total number of people seen in the cars?

   e What is the mean number of people per car?

2 Halima counts the number of peas in pea pods.

Her results are shown in the frequency table.

Number of peas in pod	Frequency
3	2
4	7
5	15
6	18
7	6
8	2

   a How many pods are shown in the table?

   b How many pods had 7 peas in?

   c Altogether, how many peas were in pods with 7 peas in?

   d What is the total number of peas in the pods?

   e What is the mean number of peas per pod?

   f What is the modal number of peas per pod?

3 Faith opened some packs containing drawing pins.

She counted the number of drawing pins in each pack.

Her results are shown in the frequency table.

Number of drawing pins in pack	Frequency
38	6
39	8
40	25
41	10
42	11

   a How many packs did Faith open?

   b Calculate the total number of drawing pins in all the packs.

   c What is the mean number of pins per pack?

4 A class of students is doing a memory test.

Objects are placed on a tray. Students have 20 seconds to look at the objects.

They then list as many objects as they can remember.

The results of the memory test are shown in the frequency table.

Number of objects remembered	Frequency
14	2
15	4
16	7
17	10
18	3
19	2

a How many people are in the class altogether?

b What is the total number of objects remembered?

c What is the mean number of objects remembered per person?

5 The frequency table shows the number of eggs in bird nests.

Find the mean number of eggs per nest.

Number of eggs in nest	Frequency
0	3
1	8
2	18
3	11

6 The frequency table shows the number of people living in houses in a particular street.

Manju says that on average more than 3 people live in each house.

Is Manju correct? Show how you decide.

Find the mean number of people living in each house to help answer the question.

Number of people in house	Frequency
0	1
1	3
2	5
3	21
4	18
5	2

7 Two groups of students, Group A and Group B, are estimating the size of an angle.

Group A estimate	Frequency
31	1
32	2
33	2
34	5
35	6
36	5
37	5
38	2

Group B estimate	Frequency
31	3
32	2
33	6
34	3
35	4
36	5
37	2
38	3

The actual size of the angle is 34°. On average, which group is closer with their estimates?

6.4 Comparing two distributions

Comparing average values

An average value is a single value used to represent a set of values (or distribution). So, when you have two distributions to compare it is much easier to make a comparison using an average value for each one.

The mode, mean and median are used as average values.

When you make comparisons it is possible to get different conclusions depending on which kind of average value you use.

Worked example 1

Pedro counted the number of cars passing him every 30 seconds on Road A.

Rodriguez did the same on Road B.

Pedro recorded the following: 5, 12, 2, 6, 4, 11, 16, 10, 8, 4.

Rodriguez recorded the following: 3, 5, 6, 8, 4, 7, 18, 12, 9.

Which of the two roads is busier?

Pedro recorded 10 values and Rodriguez recorded 9 values.

Because the number of values are different, you can not compare the two roads by just adding the values.

You must use an average to compare the two roads.

In this example, the question does not say which average to use.

There is no mode for Road B so finding the mode for Road A would not help answer the question.

You could use

a the mean or

b the median.

a Using the mean

The mean for Road A $= \dfrac{5 + 12 + 2 + 6 + 4 + 11 + 16 + 10 + 8 + 4}{10} = \dfrac{78}{10} = 7.8$

The mean for Road B $= \dfrac{3 + 5 + 6 + 8 + 4 + 7 + 18 + 12 + 9}{9} = \dfrac{72}{9} = 8$

So, on average, Road B is busier because it has a higher mean.

b Using the median

For Road A:

   2, 4, 4, 5, 6, 8, 10, 11, 12, 16

The median $= \dfrac{6 + 8}{2} = 7$

> To find the median, list the values in order.

For Road B:

   3, 4, 5, 6, 7, 8, 9, 12, 18

The median $= 7$

So, on average, both roads are as busy as each other because they have the same median.

Worked example 2

Two groups of students record their shoe sizes.

Group A records 5, 5, 5, 6, 6, 6, 7, 7, 7

Group B records 3, 4, 5, 6, 7, 8, 9

Which group wears the smallest shoes?

As in Worked example 1, the question does not tell you which average to use.

Group B has no mode. This leaves the mean and the median.

Group A: the mean is 6 and the median is 6.

Group B: the mean is 6 and the median is 6.

On average, both groups take the same shoe size.

Comparing values for a measure of spread

In Worked example 2 the average values are the same, although the shoe sizes for the two groups are quite different. A measure of spread gives more information about the distribution of values.

The range is a measure of spread.

Worked example 3

Use the range of the distributions to compare the shoe sizes for Groups A and B in Worked example 2.

Group A: range = 7 − 5 = 2

Remember: The range = largest − smallest.

Group B: range = 9 − 3 = 6

Group B has a larger range than Group A.

This tells us that the shoe sizes for Group B are more spread out than the shoe sizes for Group A.

Worked example 4

The time, in seconds, taken to answer a phone by two telephone receptionists is recorded.

Receptionist A: 3, 5, 4, 7, 6, 3, 4, 4

Receptionist B: 2, 3, 9, 7, 3

a Give a reason why A is a better receptionist.

b Give a reason why B is a better receptionist.

First find average values and the range.

For receptionist A:

The mean time is $\dfrac{3 + 5 + 4 + 7 + 6 + 3 + 4 + 4}{8} = \dfrac{36}{8} = 4.5$ seconds

The median time is 4 seconds.

The range is $7 - 3 = 4$ seconds

For receptionist B:

The mean time is $\dfrac{2 + 3 + 9 + 7 + 3}{5} = \dfrac{24}{5} = 4.8$ seconds

The median time is 3 seconds.

The range is $9 - 2 = 7$ seconds

a Using the mean value: Receptionist A is a better receptionist as, on average, they are faster because their mean value is lower. Their range is lower, which means they are also more consistent (their times are less spread out).

b Using the median value: Receptionist B is a better receptionist as, on average, they are faster because their median value is lower.

Exercise 6.4

1 The table shows some information about three flocks of sheep.

a Which flock generally has the greater mass? Explain your answer.

b In which flock are the masses spread out most? Explain your answer.

Flock	Mean	Range
Flock A	130 kg	43 kg
Flock B	135 kg	32 kg
Flock C	115 kg	48 kg

c In which flock are the masses least spread out? Explain your answer.

2 Here are the heights of two netball teams.

Team A: 156 cm, 158 cm, 169 cm, 174 cm, 149 cm, 155 cm, 166 cm

Team B: 162 cm, 153 cm, 172 cm, 152 cm, 171 cm, 155 cm, 159 cm

a Find the median height of each team.

b What does the median tell you about the heights of the two teams?

c Find the range of heights of each team.

d What does the range tell you about the heights of the teams?

3 Four machines fill bags with crisps. Each bag is supposed to have a mass of 32 g.

The mass of 8 bags from each machine is recorded.

Machine A: 31 g, 33 g, 29 g, 32 g, 34 g, 33 g, 32 g, 35 g

Machine B: 32 g, 29 g, 33 g, 27 g, 34 g, 36 g, 34 g, 31 g

Machine C: 28 g, 37 g, 32 g, 33 g, 26 g, 29 g, 31 g, 34 g

Machine D: 29 g, 31 g, 34 g, 33 g, 30 g, 35 g, 34 g, 33 g

  a Find the mean and range for each machine.

  b Which machine generally under filled the bags?

  c Which machine was most variable in filling the bags?

4 A weather station records the rainfall, in mm, each month.

Year	Jan	Feb	Mar	Apr	May	Jun	Jul	Aug	Sep	Oct	Nov	Dec
1910	111	126	50	95	72	70	97	140	27	89	128	142
1960	125	99	51	69	46	59	111	118	102	146	147	126
2010	80	75	79	48	39	39	108	98	114	101	123	48

  a Find the mean rainfall per month for each year.

  b Find the range of the amount of rainfall for each year.

  c Which year was generally the wettest?

  d In which year did the amount of rain per month vary most?

  e Write a few sentences comparing the rainfall for the three years.

5 A student lists the number of words in each sentence on the first page of two books.

Book 1: 5, 15, 16, 32, 6, 17, 19, 30, 13, 21, 24, 32, 14

Book 2: 12, 16, 6, 18, 8, 8, 10, 22

Use the mean and range to compare the first page of each book.

6 The frequency tables show the numbers of fish caught in one day by people in two fishing clubs.

Rod's club	
Number of fish	**Frequency**
0	2
1	5
2	8
3	2
4	4
5	3

Annette's club	
Number of fish	**Frequency**
0	6
1	4
2	3
3	4
4	4
5	6

  a Use the mean and range to compare the number of fish caught by the two clubs.

  b Use the median to compare the number of fish caught by the two clubs.

  c Compare your answers to part **a** and part **b**.

Review

1 Find the mode, median, mean and range for each of the following:

a 1, 2, 2, 3, 4, 5, 5, 5, 9

b 10, 12, 18, 19, 21

c 30, 30, 32, 34, 35, 36, 38, 38, 39, 45

2 The frequency table shows the number of people living in houses in a street.

Number of people in house	Frequency
0	1
1	3
2	6
3	7
4	5
5	2

a How many houses have 3 people living in them?

b How many houses are in the street?

c How many people live in the street altogether?

d What is the mean number of people per house?

3 The frequency tables show the number of books read last month by students in two classes.

Class A		Class B	
Number of books	Frequency	Number of books	Frequency
0	2	0	1
1	3	1	1
2	6	2	9
3	5	3	5
4	5	4	5
5	5	5	3
6	4	6	6

Use the mean and range to compare the number of books read by the two classes.

7 Length, mass and capacity

Made to measure

Historically, people have used many different things as a basis for measurement.

Early humans used the span of the human hand or the cubit, based on the length from the elbow to the finger tip, as units for measurement. The foot, based on the length of a human foot, is still in use.

For measuring mass, some systems were based on the mass of a grain of wheat.

Nowadays, we use a standardised system of units, which is based on the metric system first introduced in France in the 1790s.

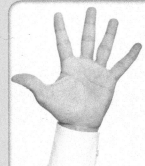

7.1 Length

The metre (m) is a unit of length.

You use a ruler, a metre rule or a measuring tape to measure lengths.

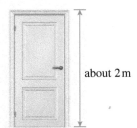

about 2 m

about 4.5 m

The height of a door is about 2 metres.   The length of a car is about 4.5 metres.

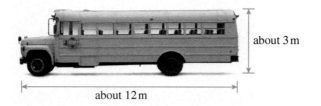

about 3 m

about 12 m

A bus is about 12 metres long and about 3 metres high.

For shorter lengths, you use the centimetre (cm) or the millimetre (mm).

A centimetre is equal to $\frac{1}{100}$ of a metre. A millimetre is equal to $\frac{1}{1000}$ of a metre.

about 15 cm

This pen is about 15 cm or 150 mm long.

> In metric units 'centi' always means $\frac{1}{100}$ of the unit.
>
> 'Milli' always means $\frac{1}{1000}$ of the unit.

You need to remember the following conversion factors:

1000 millimetres = 1 metre

100 centimetres = 1 metre

10 millimetres = 1 centimetre

For longer lengths, you use the kilometre (km).

1 kilometre = 1000 metres

> In metric units 'kilo' always means 1000 of the unit.

Worked example

Convert the following:

a 4.6 km into metres   **b** 5.9 m into centimetres   **c** 3450 mm into metres.

a 1 km = 1000 m

  4.6 km = 4.6 × 1000 m

  4.6 km = 4600 m

> When changing from a bigger unit to a smaller unit, you **multiply** by the conversion factor.

b 1 m = 100 cm

5.9 m = 5.9 × 100 cm ●————

5.9 m = 590 cm

> When changing from a bigger unit to a smaller unit, you **multiply** by the conversion factor.

c 1 m = 1000 mm

3450 mm = 3450 ÷ 1000 m ●————

3450 mm = 3.45 m

> When changing from a smaller unit to a bigger unit, you **divide** by the conversion factor.

Exercise 7.1

1 Estimate:

   a the length of this pencil case in centimetres

   b the height of the man in metres

   c the height of the tree in metres

   d the thickness of this pencil in millimetres.

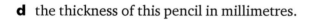

2 What unit would you use to measure:

   a the height of a table   **b** the thickness of this book   **c** the length of a football field?

3 Convert these lengths to metres.

   a 6 km   **b** 2.9 km   **c** 12.5 km   **d** 0.5 km   **e** 0.067 km

   f 300 cm   **g** 450 cm   **h** 58 cm   **i** 1250 cm   **j** 12.5 cm

   k 6.8 cm   **l** 4000 mm   **m** 8650 mm   **n** 35 mm   **o** 8 mm

4 Convert these lengths to centimetres.

   a 4 m   **b** 9.6 m   **c** 15.7 m   **d** 0.65 m   **e** 3.87 m

   f 0.54 m   **g** 0.06 m   **h** 70 mm   **i** 17 mm   **j** 98 mm

   k 134 mm   **l** 6 mm   **m** 398 mm   **n** 0.7 mm

5 Convert these lengths to millimetres.

   a 2 m   **b** 1.8 m   **c** 60.4 m   **d** 0.85 m   **e** 2.3 cm

   f 7.5 cm   **g** 0.8 cm   **h** 47 cm   **i** 0.05 cm

6 Convert these lengths to kilometres.

   a 1500 m   **b** 8530 m   **c** 12 500 m   **d** 650 m   **e** 354 m

   f 75 m   **g** 36.8 m   **h** 12.8 m   **i** 8 m

7 A swimmer completes 30 lengths of a 50 metre pool.

   a How many metres does she swim?

   b How many kilometres does she swim?

   c Her friend wants to swim 5 kilometres. How many lengths is this?

Be careful with different units in a question.

8 This Australian postage stamp measures 26 mm by 31.5 mm.

What are these lengths in centimetres?

9 An athlete completes 25 laps of a 400 metre track. How many kilometres has he run?

10 The base of a farmer's cart is 1.3 m above the ground.

The farmer is stacking bales of straw on to the cart.

Each bale is 45 cm deep.

He stacks 6 layers of bales on the cart.

A bridge is 4.1 m high.

Will the cart fit under the bridge?

45 cm

1.3 m

11 Helen is training for a triathlon. She needs to swim, then cycle and then run.

She starts in a swimming pool which is 25 metres long. She swims 36 lengths of the pool.

Next she cycles along roads for 12.5 kilometres.

Finally she arrives at a 400 metre athletics track. She runs around the track 14 times.

She tells her parents that the total distance is 20 kilometres.

Is she right? You must show the working.

12 Joseph is organising a fund-raising activity.

He is trying to create a one kilometre line of coins.

Each coin is 21 mm across.

Work out how many coins are needed to cover 1 kilometre.

Change the kilometre into metres and then change that into millimetres.

7.2 Mass

You use scales to measure mass.

The gram (g) is a small unit of mass.

Here are some examples of the mass of objects.

A pen has a mass of about 15 g.	An egg has a mass of about 50 g.	An orange has a mass of about 100 g.

For heavier objects, you use the kilogram (kg).

1 kilogram = 1000 grams •————————————— Remember that kilo means 1000.

A small pineapple has a mass of about 1 kg.   A medium adult dog has a mass of about 10 kg.

For even heavier objects, you use a tonne (t).

1 tonne = 1000 kilograms

A car has a mass of about 2 tonnes.

Worked example 1

What unit would you use to measure the mass of:

a an elephant   **b** this book?

a An adult elephant is very heavy so you would probably use tonnes, although you may choose to use kilograms for a young elephant.

b This book has a mass of less than one kilogram, so you would use grams.

Worked example 2

Convert the following:

a 3.4 kilograms to grams **b** 7650 grams to kilograms **c** 13 500 kilograms to tonnes.

a 1 kg = 1000 g

3.4 kg = 3.4 × 1000 g

3.4 kg = 3400 g

b 1 kg = 1000 g

7650 g = 7650 ÷ 1000 kg

7650 g = 7.65 kg

c 1 t = 1000 kg

13 500 kg = 13 500 ÷ 1000 t

13 500 kg = 13.5 t

> Remember, when converting from a bigger unit to a smaller unit you multiply by the conversion factor and when converting from a smaller unit to a bigger unit you divide by the conversion factor.

Exercise 7.2

1 What unit would you use to measure the mass of:

  a a dog   **b** a bus   **c** a spoonful of sugar   **d** a horse?

2 Jane found the mass of several things but she forgot to write down the units.

Complete these statements with the appropriate unit.

  a The mouse on my computer has a mass of 80 …

  b A discus used in the Olympic Games has a mass of 2 …

  c My science textbook has a mass of 1.2 …

  d My father's motorbike has a mass of 345 …

3 Convert the following units.

  a 5 kilograms to grams   **b** 8.9 kilograms to grams   **c** 0.7 kilograms to grams

  d 3500 grams to kilograms   **e** 13 870 grams to kilograms   **f** 250 grams to kilograms

  g 5000 kilograms to tonnes   **h** 2750 kilograms to tonnes   **i** 2.5 tonnes to kilograms

  j 45 tonnes to kilograms

4 Put these in order of mass, starting with the smallest.

13 600 g, 0.03 t, 2.8 kg, 43 kg

5 Seven people want to use a lift. Their masses are 60.7 kg, 57.7 kg, 70.8 kg, 81.7 kg, 65.9 kg, 80.4 kg and 74.1 kg.

> Find the total mass in kilograms. Remember that 1 tonne = 1000 kg.

The maximum load allowed in the lift is 0.5 tonnes. Are they safe to travel in the lift?

6 Three brothers have bought presents to send to their grandparents.

The delivery company will take parcels with mass of up to 3 kilograms.

The masses of their presents are 1.35 kg, 0.75 kg and 240 g.

The box that they pack them in has a mass of 30 g.

They want to buy one more present to put in the box.

What is the biggest mass that the extra present can have?

7 A milligram (mg) is $\frac{1}{1000}$ of a gram so 1000 mg = 1 g.

A box of sweets contains 480 sweets. Each sweet has a mass of 150 milligrams.

Work out the total mass of the sweets in grams.

7.3 Capacity

Capacity is a measure of the amount that a container can hold.

The **litre (l)** is used to measure capacity.

You also measure amounts of liquids in litres.

You can use a measuring jug or measuring spoons to measure an amount of liquid.

For smaller amounts you use the **millilitre (ml)**.

1 litre = 1000 millilitres •————————— Remember that milli means $\frac{1}{1000}$

Here are some examples of capacity.

The bucket holds 5 l.　　The cup holds 250 ml.　　The carton of juice holds 1 l.　　The medicine spoon holds 5 ml.

Worked example

Convert the following:

a 5.7 litres to millilitres　　**b** 8500 millilitres to litres　　**c** 135 millilitres to litres.

a 1 l = 1000 ml　　　　　**b** 1 l = 1000 ml'　　　　**c** 1 l = 1000 ml

　5.7 l = 5.7 × 1000 ml　　　8500 ml = 8500 ÷ 1000 l　　135 ml = 135 ÷ 1000 l

　5.7 l = 5700 ml　　　　　8500 ml = 8.5 l　　　　135 ml = 0.135 l

Exercise 7.3

1 What unit would you use to measure:

    a the capacity of a tea cup   **b** the amount of water in a swimming pool?

2 Convert the following units.

    a 5 litres to millilitres   **b** 6.8 litres to millilitres

    c 12 litres to millilitres   **d** 0.7 litres to millilitres

    e 8750 millilitres to litres   **f** 13 500 millilitres to litres

    g 800 millilitres to litres   **h** 60 millilitres to litres

3 Put these in order, starting with the smallest.

1.2 l, 1340 ml, 0.765 l, 2345 ml, 750 ml

4 Alma has a bottle that contains 1.5 l of milk.

She needs 350 ml of milk to make a hot drink.

She needs 0.7 l to make a dessert.

She needs 500 ml to make her son a drink.

Is there enough milk in the bottle?

> Convert everything to millilitres.

5 Shani has a 2 litre bottle of milk. She is using it to make some desserts.

Each dessert needs 120 ml of milk.

How many desserts can Shani make?

6 One litre of pure water has a mass of one kilogram.

    a What is the mass of one millilitre of water?

    b How many tonnes of water are there in a swimming pool that holds 400 000 l?

> Remember that
> 1 litre = 1000 millilitres and
> 1 tonne = 1000 kilograms.

7.4 Reading from scales

To find length, mass or capacity, you often have to read from a scale.

A scale may be straight, like the one on the left, or it may be curved, like the one on the right.

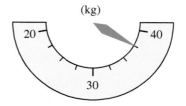

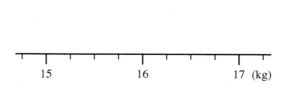

In either case, the method for reading the scale is the same. Look at this scale.

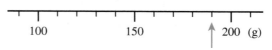

Look at the values marked on the scale and work out the difference between them.

This difference is called an interval.

In this case there is 50 g between the values.

Look at how many spaces the interval is divided into.

In this case there are 5 spaces.

$50 \div 5 = 10$ so each space represents 10 g.

The reading shown is 190 g.

> Do not count the lines in the intervals – count the spaces between the lines.

Worked example

a Read the value shown on this scale.　　**b** Estimate the value shown on this scale.

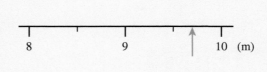

a There are 10 grams between each value. There are 10 divisions in each interval.

Each division represents 1 g.

The reading is at 24 g

b There is 1 metre between each value and 2 divisions in each interval.

Each division represents 0.5 m.

The reading is above 9.5 m.

It is nearer to 9.5 than to 10.

9.7 m is a good estimate.

Digital scales

The scales on newer, electronic, measuring equipment are often digital.

They can be read directly as a number.

There is no need to work out the intervals and the divisions.

These scales show the reading 2.034 kg very clearly.

Exercise 7.4

1 Write down the readings from each of these scales.

a

(kg)

20   40

30

b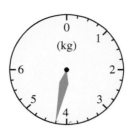

c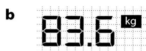

5   6   7 (cm)

d

e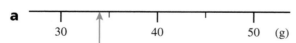

30   40   50 (g)

2 Read the mass shown on each of these digital scales.

a kg 1.090

b 83.6 kg

3 How much liquid is there in each of these measuring jugs? Give your answers in millilitres.

a

(l)

3

2

1

b

(ml) 200

100

4 Estimate the readings from each of these scales.

a

30   40   50 (g)

b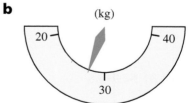

(kg)

20   40

30

c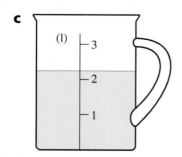

(l)

3

2

1

d

(ml) 200

100

e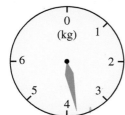

f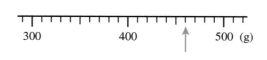

5 These scales show the masses of three packages.

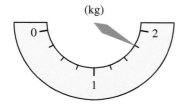

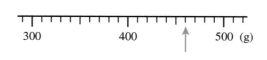

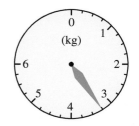

a Write down the mass of each parcel in kilograms.

Here is a chart showing the price for posting parcels.

Use the chart to work out:

b the cost of posting each one separately

c the cost of posting them all together in one big parcel

d how much cheaper it is to post them together as one parcel.

Mass of the parcel	Cost
0 – 1 kg	$6.90
1 – 2 kg	$10.85
2 – 5 kg	$18.42
5 – 10 kg	$25.95

6 Jess and Walter are measuring the temperature in each of three rooms.

The three readings are shown here. They are all in degrees Celsius.

A Bedroom

B Bathroom

C Living room

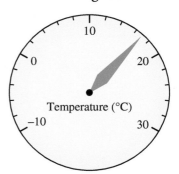

a How much higher is the temperature in the bathroom than in the bedroom?

b Which room has the highest temperature? Write down the temperature in that room.

c Which room has the lowest temperature? Write down the temperature in that room.

Review

1 Copy and complete these conversions.

    a 3.4 km = ? m   **b** 543 cm = ? m   **c** 230 mm = ? cm   **d** 1.2 m = ? mm

    e 8650 mm = ? m   **f** 7390 m = ? km   **g** 98 cm = ? mm   **h** 4.6 m = ? cm

2 Copy and complete these conversions.

    a 6.9 kg = ? g   **b** 8500 kg = ? t   **c** 2.5 t = ? kg   **d** 7560 g = ? kg

3 Copy and complete these conversions.

    a 1900 ml = ? l   **b** 7.5 l = ? ml   **c** 12 l = ? ml   **d** 750 ml = ? l

4 Put these lengths in order, starting with the smallest.

    500 m, 12 000 cm, 0.4 km, 800 mm

5 A farmer has these five sacks.

    a What is the total mass in kilograms?

    b What is the total mass in tonnes?

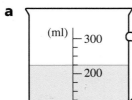

120 kg   25 kg   75 kg   200 kg   80 kg

6 Tom has 0.6 litres of paint.

    Polly has five pots, each containing 150 millilitres of paint.

    Who has more paint, Tom or Polly?

7 What would be the most appropriate metric unit for measuring these:

    a the length of a finger nail   **b** the mass of a chicken

    c the height of a building   **d** the capacity of a watering can?

8 Write down the measurements shown on these scales.

    a   **b**   **c**

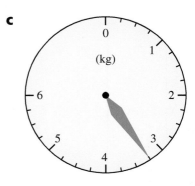

9 Estimate the measurements shown on these scales.

    a   **b**

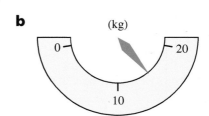

8 Equations

The equal sign

Mathematicians use many different mathematical symbols.

One of the most commonly used symbols is the equal sign.

How many equal signs (=) can you find in this picture?

In 1557, a mathematician called Robert Recorde invented the equal sign that we use today. He first used the symbol in a book called *The Whetstone of Witte*.

Robert Recorde was a Welshman and he wrote many mathematical textbooks in his lifetime.

8.1 Equations with one operation

$x + 2$ is an expression.

$x + 2 = 5$ is an **equation**. ●————

> An equation contains a variable and an equal sign.

To find the value of x in the equation $x + 2 = 5$ you must **solve** the equation.

You can think of the equation as a balance.

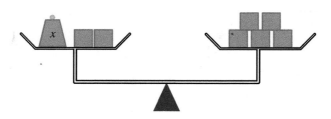

To find the value of x, you need a way of leaving x on its own.

You must take the same amount from both sides to keep the scales balanced.

So take 2 from both sides.

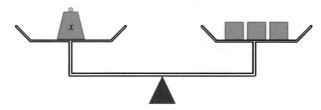

The solution of the equation $x + 2 = 5$ is $x = 3$

You do not need to use a pair of scales each time you solve an equation.

You need to know the following rules for solving equations:

- You can add the same number to both sides.
- You can subtract the same number from both sides.
- You can multiply both sides by the same number.
- You can divide both sides by the same number.

> Remember to do the same thing to both sides.

To solve the equation without the pair of scales:

$$x + 2 = 5 \qquad \text{subtract 2 from both sides}$$
$$x + 2 - 2 = 5 - 2$$
$$x = 3$$

To solve equations you use inverse operations.

In the example above the inverse of the operation 'add 2' is 'subtract 2'.

You should always check your answers by substituting back into the original equation.

Worked example

Solve these equations.

a $x + 6 = 9$ **b** $x - 5 = 13$ **c** $3x = 21$ **d** $\frac{x}{4} = 6$

a $x + 6 = 9$ subtract 6 from both sides   •   The inverse of 'add 6' is 'subtract 6'.
$$x = 9 - 6$$
$$x = 3$$
Check: $3 + 6 = 9$ ✓

b $x - 5 = 13$ add 5 to both sides   •   The inverse of 'subtract 5' is 'add 5'.
$$x = 13 + 5$$
$$x = 18$$
Check: $18 - 5 = 13$ ✓

c $3x = 21$ divide both sides by 3 •——— The inverse of 'multiply by 3' is 'divide by 3'.

$x = \frac{21}{3}$ •

$x = 7$ ———— $21 \div 3$ is the same as $\frac{21}{3}$

Check: $3 \times 7 = 21$ ✓

d $\frac{x}{4} = 6$ multiply both sides by 4 •——— The inverse of 'divide by 4' is 'multiply by 4'.

$x = 4 \times 6$

$x = 24$

Check: $24 \div 4 = 6$ ✓

Exercise 8.1

Solve these equations.

1 $x + 5 = 11$ **2** $x - 4 = 9$ **3** $x - 1 = 7$ **4** $3 + x = 15$

5 $10 = x + 2$ **6** $8 = x - 4$ **7** $18 = 5 + x$ **8** $10 = x + 14$

$10 = x + 2$ is the same as $x + 2 = 10$

9 $x - 3 = 20$ **10** $17 = x + 12$ **11** $54 = x - 36$ **12** $7 + x = -6$

13 $2x = 12$ **14** $3x = 12$ **15** $4x = 32$ **16** $20 = 5x$

17 $108 = 6x$ **18** $2x = 1$ **19** $9 = 2x$ **20** $4x = 0$

21 $\frac{x}{2} = 6$ **22** $\frac{x}{4} = 7$ **23** $32 = \frac{x}{8}$ **24** $1\frac{1}{2} = \frac{x}{10}$

25 For each part of the question you must:

     i write down an equation

     ii solve the equation.

   a Mia thinks of a number and adds 7.
The answer is 19.

What was the number?

   b Omar thinks of a number and multiplies it by 2.
The answer is 36.

What was the number?

   c Sarif thinks of a number and subtracts 11.
The answer is 27.

What was the number?

   d Sonita thinks of a number and divides by 6.
The answer is 9.

What was the number?

8.2 Equations with two operations

In section 8.1 you learned how to solve equations with one operation.

In this section you will learn how to solve equations with two operations.

Worked example 1

Solve these equations.

a $3x + 9 = 21$ **b** $\dfrac{x - 6}{4} = 5$ **c** $17 - 3x = 5$

a $3x + 9 = 21$ subtract 9 from both sides

   $3x = 12$ divide both sides by 3

   $x = \dfrac{12}{3}$

   $x = 4$

Check: $3 \times 4 + 9 = 12 + 9 = 21$ ✓

b $\dfrac{x - 6}{4} = 5$ multiply both sides by 4

   $x - 6 = 20$ add 6 to both sides

   $x = 26$

Check: $\dfrac{26 - 6}{4} = \dfrac{20}{4} = 5$ ✓

c $17 - 3x = 5$ make the x's positive by adding $3x$ to both sides

   $17 = 5 + 3x$ subtract 5 from both sides

   $12 = 3x$ divide both sides by 3

   $x = 4$

Check: $17 - 3 \times 4 = 17 - 12 = 5$ ✓

> Sometimes your answer will not be an integer (whole number).
>
> When your answer is not an integer it is best to write your answer as a fraction.

Worked example 2

Solve the equation $5x + 1 = 12$

   $5x + 1 = 12$ subtract 1 from both sides

   $5x = 11$ divide both sides by 5

   $x = \dfrac{11}{5}$ change $\frac{11}{5}$ to a mixed number

   $x = 2\frac{1}{5}$

Check: $5 \times 2\frac{1}{5} + 1 = 11 + 1 = 12$ ✓

Exercise 8.2

Solve these equations.

1 $2x + 3 = 11$

2 $3x - 4 = 14$

3 $5x + 2 = 17$

4 $2x - 1 = 9$

5 $4x - 5 = 27$

6 $6x + 3 = 57$

7 $7x - 5 = 16$

8 $3 + 2x = 15$

9 $9 + 2x = 23$

10 $16 = 2x - 4$

11 $27 = 6x + 3$

12 $7 + 3x = 16$

13 $7 = 2x - 15$

14 $5 = 2x + 5$

15 $4x - 12 = -4$

16 $-13 = 5x - 28$

17 $\dfrac{x + 2}{5} = 6$

18 $\dfrac{x - 3}{2} = 15$

19 $6 = \dfrac{3 + x}{4}$

20 $2 = \dfrac{x - 8}{3}$

21 $4 = \dfrac{x - 2}{5}$

22 $10 = \dfrac{x + 7}{3}$

23 $1 = \dfrac{6 + x}{3}$

24 $\dfrac{x + 15}{2} = 3$

25 $\dfrac{x}{2} + 1 = 5$

26 $7 + \dfrac{x}{3} = 10$

27 $8 = \dfrac{x}{5} + 2$

28 $\dfrac{x}{4} - 8 = -3$

Solve these equations. Give your answers as fractions.

29 $2x + 1 = 10$

30 $3x - 5 = 5$

31 $3x + 2 = 10$

32 $4x - 1 = 5$

33 $2x - 5 = 6$

34 $3x + 3 = 4$

35 $5x - 2 = 7$

36 $3 + 2x = 8$

37 $5 + 2x = 18$

38 $16 = 10x - 2$

39 $17 = 4x + 3$

40 $9 + 3x = 20$

41 $8 = 2x - 15$

42 $16 = 2x + 3$

43 $4x - 10 = -4$

44 $-13 = 5x - 17$

45 $2x - 3 = 0$

46 $0 = 4x - 15$

47 $10x + 10 = 11$

48 $7x - 5 = 8$

49 $20 - 2x = 14$

50 $23 - 5x = 13$

51 $42 - 6x = 18$

52 $100 - 7x = 44$

53 For each part of the question you must:

    i write down an equation

    ii solve the equation.

  a Petra thinks of a number, multiplies it by 2 and then adds 5. The answer is 17.
What was the number?

  b Tareq thinks of a number, multiplies it by 3 and then subtracts from 38. The answer is 23.
What was the number?

  c Sarif thinks of a number, divides it by 2 and then subtracts 11. The answer is 4.
What was the number?

54

6cm

144 cm²

A square's area is side length times side length.

x cm

The perimeter of the rectangle is equal to the perimeter of the square.

Find the value of x.

55 Solve these equations. Simplify by collecting like terms first.

  a $2x + 3 + 4x = 33$

  b $3x + 4 + 2x + 1 = 45$

  c $9x - 3 - 7x + 5 = 10$

  d $20 = 2x - 1 + x$

8.3 Equations with brackets

This section is an extension section.

It combines solving equations with the work that you did in Chapter 2 on expanding brackets.

> **Worked example**
>
> Solve the equation $6(x - 2) = 42$
>
> | $6(x - 2) = 42$ | expand the brackets |
> |---|---|
> | $6x - 12 = 42$ | add 12 to both sides |
> | $6x = 54$ | divide both sides by 6 |
> | $x = 9$ | |
>
> **Check:** $6 \times (9 - 2) = 6 \times 7 = 42$ ✓

Exercise 8.3

Multiply out the brackets and then solve these equations.

1 $3(x + 4) = 18$ **2** $5(x + 2) = 25$

3 $4(x + 3) = 44$ **4** $3(x - 5) = 15$

5 $6(x - 1) = 30$ **6** $8(x - 5) = 0$

7 $14 = 2(x - 8)$ **8** $36 = 9(x - 3)$

9 $2(3x + 4) = 20$ **10** $4(2x - 5) = 28$

11 $30 = 5(2x - 8)$ **12** $68 = 4(5x + 2)$

13 $3(x + 1) = 8$ **14** $15 = 2(x + 4)$

15 $5(2x - 1) = 7$ **16** $1 = 3(2x - 3)$

17 $0 = 4(3x - 2)$ **18** $2 + 3(x + 4) = 29$

19 $5 + 3(x - 2) = 23$ **20** $3(x + 5) - 4 = 41$

21 $4(x - 7) + 3 = 15$ **22** $28 = 2 + 2(3x + 4)$

23 $67 = 3(4x - 3) - 8x$ **24** $34 = x + 4(3x + 2)$

25 $3(x + 4) + 2(x - 5) = 32$ **26** $5(2x + 1) + 3(x - 4) = 71$

27 $4(x - 3) + 2(3 - x) = 44$ **28** $15 = 2(3 - 2x) + 5(x - 4)$

8.4 Using equations to solve problems

Equations can be used to solve problems.

Worked example 1

The diagram shows three rods.

a Write down an equation connecting the three lengths.

b Solve your equation to find the value of x.

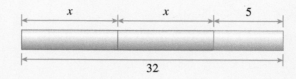

a The three lengths add up to 32.

$x + x + 5 = 32$ collect like terms

$2x + 5 = 32$

b $2x + 5 = 32$ subtract 5 from both sides

$2x = 27$ divide both sides by 2

$x = 13.5$

Worked example 2

a Write down an equation connecting the three angles of the triangle.

b Solve your equation to find the value of x.

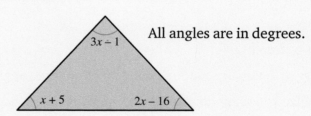

All angles are in degrees.

a The three angles of a triangle add up to $180°$.

$3x - 1 + x + 5 + 2x - 16 = 180$ collect like terms

$6x - 12 = 180$

b $6x - 12 = 180$ add 12 to both sides

$6x = 192$ divide both sides by 6

$x = 32$

Exercise 8.4

1 The diagram shows three rods.

   a Write down an equation connecting the three lengths.

   b Solve your equation to find the value of x.

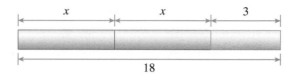

2 The diagram shows three rods.

   a Write down an equation connecting the three lengths.

   b Solve your equation to find the value of y.

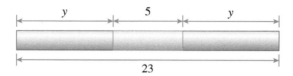

3 The diagram shows three rods.

   a Write down an equation connecting the three lengths.

   b Solve your equation to find the value of z.

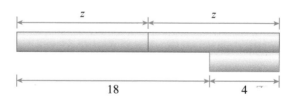

4 AB is a straight line.

   a Write down an equation connecting the three angles.

   b Solve your equation to find the value of x.

> Angles on a straight line add up to 180°.

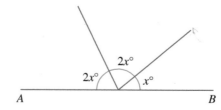

5 The sum of three consecutive integers (whole numbers) is 165.

Let the first (smallest) number be n.

$\boxed{n} + \boxed{?} + \boxed{?} = 165$

a Write down an expression for the second number.

b Write down an expression for the third number.

c Write down an equation connecting your three numbers.

d Solve the equation to find the value of n.

e Write down the three numbers.

6 a Write down an equation connecting the three angles of the triangle.

b Solve the equation to find the value of x.

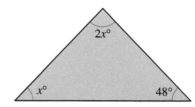

Angles in a triangle add up to 180°.

7 This is a number wall.

To find the number in each block you add the numbers in the two blocks below.

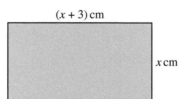

Find the value of x in this wall.

8 The perimeter of the rectangle is 42 cm.

Find the value of x.

$(x + 3)$ cm

x cm

Review

1 Solve the following.

    a $x + 4 = 11$ **b** $x - 8 = 13$ **c** $26 = 2x$ **d** $\frac{x}{3} = 7$

2 Solve the following.

    a $3x + 5 = 17$ **b** $2x - 7 = 9$ **c** $4x - 16 = 4$ **d** $\frac{x + 3}{5} = 4$

    e $\frac{x}{2} - 3 = 2$ **f** $2x + 5 = 8$ **g** $5x - 2 = 11$ **h** $\frac{x}{3} - 7 = -2$

3 Ricky thinks of a number, multiplies it by 3 and then adds 7.

The answer is 25.

What was the number?

4

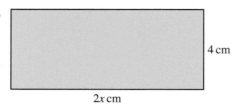

The perimeter of the rectangle is equal to the perimeter of the square.

Find the value of x.

5 Solve the following.

   a $2(x - 8) = 16$ **b** $3(x + 7) = 33$ **c** $4(x + 5) = 34$

   d $3(2x - 5) = 9$ **e** $2 + 3(x - 4) = 14$ **f** $2(5x - 3) - 6x = 22$

6 This is a number wall.

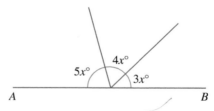

To find the number in each block you add the numbers in the two blocks below.

Find the value of y in this wall.

7 AB is a straight line.

   a Write down an equation connecting the three angles.

   b Solve the equation to find the value of x.

8 The perimeter of the rectangle is 50 cm.

Find the value of x.

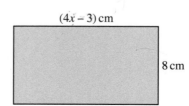

9 Shapes and geometric reasoning 2

Learning outcomes

- Start to recognise the angular connections between parallel lines, perpendicular lines and transversals.
- Use a ruler, set square and protractor to:
  - measure and draw straight lines and angles
  - draw perpendicular and parallel lines
  - construct a triangle given two sides and the included angle (SAS) or two angles and the included side (ASA)
  - construct rectangles, squares and regular polygons.
- Recognise and describe common solids and some of their properties, e.g. the number of faces, edges and vertices.

Mathematical designs

Lines, angles and polygons are found in many different places.

Here is an example of a tiling pattern using regular mathematical shapes.

This design is based on blue octagons and red squares.

These shapes are both regular polygons.

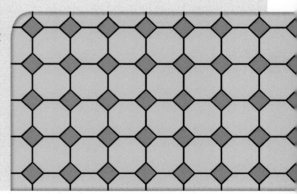

9.1 Lines and angles

Parallel lines

This is a pair of parallel lines.

They are always the same distance apart.

If they continue in either direction they never meet.

The small arrows show that the lines are parallel.

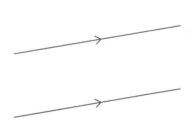

A straight line that crosses two or more parallel lines is called a transversal.

When a transversal is drawn it makes angles at both the crossing points.

This diagram shows the eight angles, four at each vertex.

Four of these angles are **acute**. These are shaded red.

All of these acute angles are equal.

Four of these angles are **obtuse**. These are shaded blue.

All of these obtuse angles are equal.

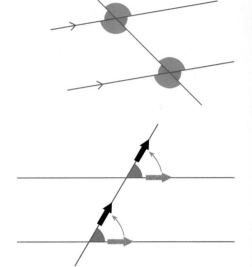

This diagram demonstrates why this is true.

The grey arrows point along the parallel lines so they point in the same direction.

The black arrows show a rotation to the other line.

The arrows still point in the same direction so they must have rotated through the same angle.

The green angles are equal.

You could do the same with the other pairs of angles.

So if you know any one angle you can work out the other angles.

Worked example 1

Find angles a, b, c, d, e, f and g.

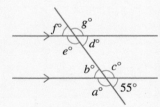

$a° = 125°$ angles on a straight line add up to 180°

$b° = 55°$ opposite angles are equal

$c° = 125°$ opposite and equal to angle a

The angles at the other vertex are the same so:

$d° = 55°$ and $f° = 55°$

$e° = 125°$ and $g° = 125°$

Worked example 2

Find angles a, b and c.

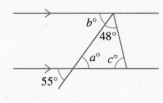

$a° = 55°$ opposite angles are equal

$b° = 55°$ the acute angles at both vertices are equal

$c° = 180° - 48° - 55°$ angles in a triangle add up to 180°

$c° = 77°$

Perpendicular lines

When two lines cross or meet at right angles they are perpendicular.

This diagram shows two parallel lines.

A third line is perpendicular to both of them.

There are eight right angles shown.

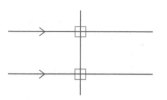

Exercise 9.1

For questions **1** to **7**, find the angles marked with letters.

1

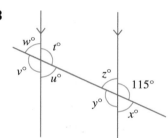

2

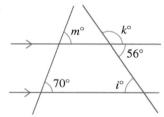

3

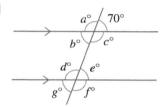

4

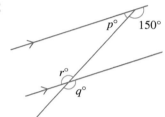

5

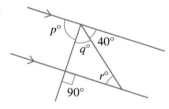

6

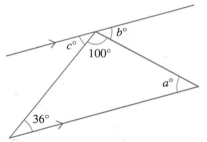

7

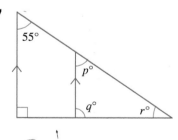

8 *ABC* is a triangle.

Angle *ABC* = 90°

Angle *ACB* = 62°

CD is parallel to *BA*.

BCE is a straight line.

Find:

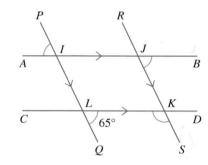

   a angle *BAC*   **b** angle *DCE*   **c** angle *ACD*.

9 *AB* and *CD* are parallel lines.

PQ and *RS* are parallel lines.

They intersect at *I*, *J*, *K* and *L*.

Angle *KLQ* = 65°

Work out:

   a angle *AIP*   **b** angle *BJK*   **c** angle *LKS*.

   d What type of quadrilateral is *IJKL*?

> It may help to look back at Chapter 3 to see the different types of quadrilateral.

10 The quadrilateral in this diagram is a kite.

Work out angle *x*.

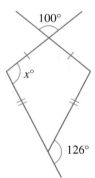

> The opposite angle in the kite is also *x*°.

11 The diagram shows a transversal crossing two parallel lines.

   a Work out angle *a*.

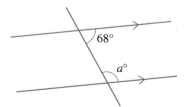

> It may help to work out the obtuse angle next to 68° first.

b Write down the sum of the two marked angles.

These are known as interior or allied angles.

When a transversal meets two parallel lines the interior or allied angles add up to 180°.

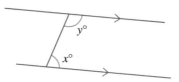

Use this fact to find the marked angles in parts **c** to **e**.

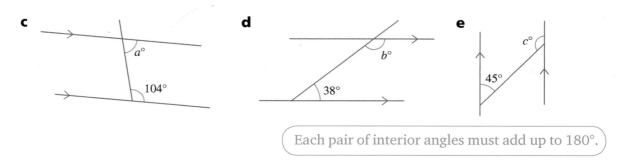

Each pair of interior angles must add up to 180°.

9.2 Measurement and construction

Measuring and drawing lengths

To measure or draw lengths you use a ruler marked in centimetres and millimetres.

In Chapter 7 you learned to work with these metric units. The ruler is marked in centimetres and each small division is one millimetre.

To use the ruler you must be sure that you start from the 0 mark. This diagram shows a red line that is 5.4 cm long.

Measuring and drawing angles

To measure or draw angles you use a protractor marked in degrees.

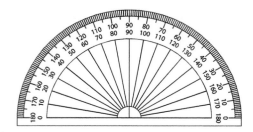

There are two scales on the protractor. One goes clockwise and the other anticlockwise.

You need to be sure that you are using the correct scale.

Worked example 1

Measure this angle.

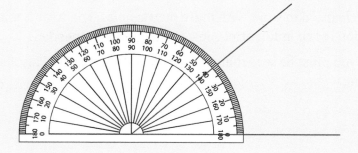

Place the protractor on the angle.

Put the central mark on the vertex as shown here.

In this case the base line touches the 0 of the outer, anticlockwise, scale.

Read from this scale – the angle is 40°.

Worked example 2

Measure this **reflex** angle.

The angle is greater than 180° but the protractor only goes up to 180°.

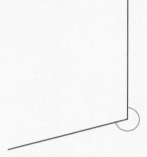

If you draw a line straight down you divide the angle into two parts.

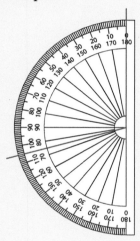

The right hand part is 180°. Measure the left hand part.

In this case you use the clockwise scale.

It is 75°, so the total reflex angle is 180 + 75 = 255°.

To **draw** a reflex angle you use the same method.

Draw a 180° angle first then work out the extra angle that is needed.

Draw the extra angle to give the required reflex angle.

This diagram shows how you would use 180° and 115° to draw an angle of 295°.

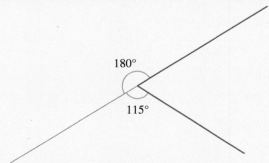

Exercise 9.2a

1 Measure these lengths.

    a _____   **b** _____

    c _____   **d** _____

2 Measure these angles.

    a   **b**

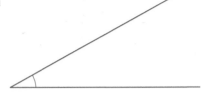
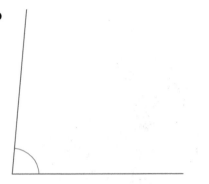

    c   **d**

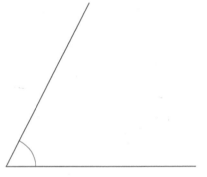
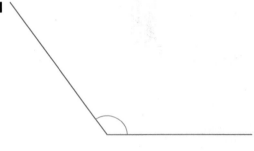

    e   **f**

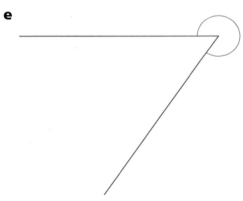
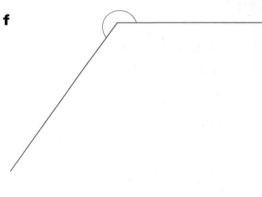

3 Draw straight lines with these lengths.

    a 6 cm   **b** 3.9 cm   **c** 7.2 cm   **d** 0.8 cm

4 Draw the following angles.

    a 50°   **b** 90°   **c** 28°   **d** 147°

    e 175°   **f** 256°   **g** 270°   **h** 318°

5 a Draw two lines 6.8 cm long with an angle of 43° between them as shown below.

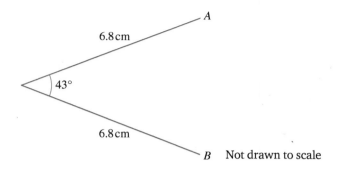

6.8 cm

A

43°

6.8 cm

B   Not drawn to scale

b Measure the distance *AB*.

In the next section you will learn how to draw perpendicular lines, parallel lines and triangles.

You need to be sure that you use your ruler, set square and protractor carefully and accurately.

Always use a sharp pencil so that the diagrams are clear.

When you are asked for a construction line, it should be a faint line that you use to help in the drawing. It should not be too dark.

Drawing perpendicular lines

Perpendicular lines are at right angles. Use your set square to draw them.

Worked example 3

Draw a line 8 cm long. Label the line *AB*.

Mark the midpoint of *AB*. Label the point *C*.

Draw a line *CD* perpendicular to *AB* such that *CD* is 5 cm.

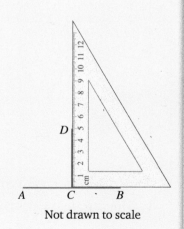

Use your ruler to draw an 8 centimetre line and label it *AB*.

Measure 4 cm along *AB* to find the midpoint and label the point *C*.

Position your set square carefully at *C*. Draw a line at right angles to *AB*.

Measure 5 cm up the line and mark the point *D*.

A   *C*   *B*

Not drawn to scale

Drawing parallel lines

Parallel lines are always the same distance apart. You can use this fact to draw them accurately.

Worked example 4

Draw two parallel lines 3 cm apart. Label them *AB* and *CD*.

First use your ruler to draw one straight line. Label this line *AB*.

Use your set square to draw a faint construction line perpendicular to *AB*.

Measure exactly 3 cm up from the base line and mark a point.

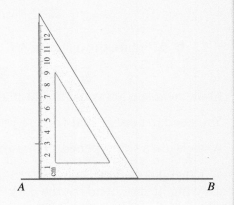

Repeat this at a point further along *AB*. You may need to turn the set square over to fit it on the line.

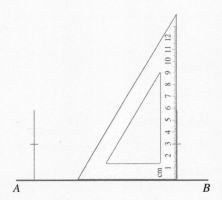

Join up the two points marked on the construction lines. Label the new line *CD*.

You have two parallel lines 3 cm apart.

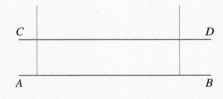

Drawing triangles

In the following example you are told the lengths of two sides and the size of the angle between them. This is sometimes described as Side-Angle-Side or simply **SAS**.

Worked example 5

Make an accurate drawing of triangle *PQR*.

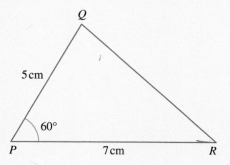

First use your ruler to draw accurately a line 7 cm long.
Label the line *PR*.

Now position your protractor at *P* and measure an angle of 60°.

Draw a mark to show 60°.

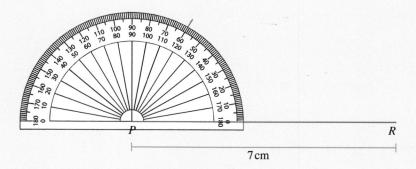

Use your ruler to draw a 5 cm line at 60° to *PR*. Label the end point *Q*.

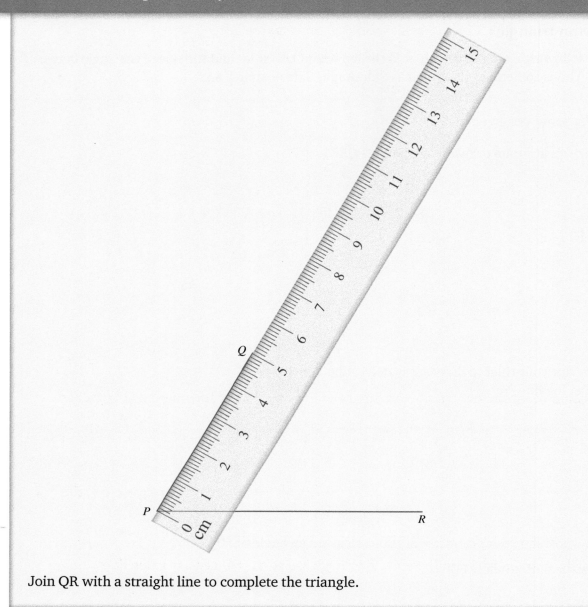

Join QR with a straight line to complete the triangle.

In the following example of a triangle you are given the sizes of two angles and the length of the side between them. This is known as Angle-Side-Angle or **ASA**.

Worked example 6

Make an accurate drawing of this triangle.

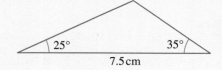

25°   35°

7.5 cm

First use your ruler to draw the base line 7.5 cm long.

Position your protractor at the end of the line and mark an angle of 25°.

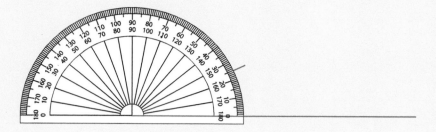

Draw a faint construction line through the marking.

Position your protractor at the other end and draw a construction line at 35° to the base.

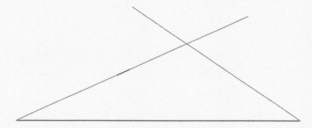

Finally, draw and label the completed triangle. Leave the faint construction lines.

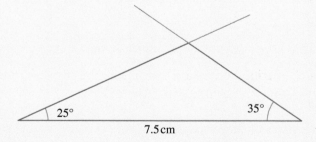

Exercise 9.2b

1 Draw two parallel lines 9 cm long and 4 cm apart.

2 Draw line *AB*, 11 cm long. Draw a parallel line 3.5 cm from *AB*. Label this line *CD*.

3 Make accurate drawings of these diagrams.

a

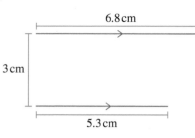

b

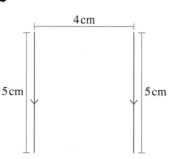

c

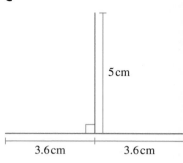

4 PQ is a straight line 8 cm long. R is a point on PQ, 3 cm from P.

   RS is perpendicular to PQ. $RS = 4$ cm.

   a Make an accurate drawing of the lines.

   b Measure the distance PS.

   If there is no diagram it may help to draw a sketch first.

5 a Make an accurate drawing of this triangle.

   b Measure length AC.

   c Measure angle BCA.

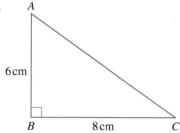

6 Make accurate drawings of these triangles.

a

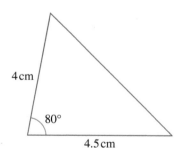

b

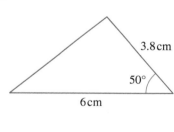

c

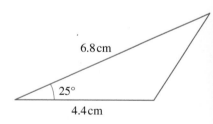

7 Make accurate drawings of these triangles.

a

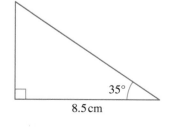

b

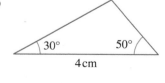

c

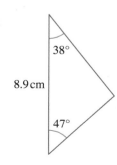

8 Make accurate drawings of these triangles.

a

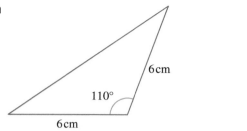

6cm

110°

6cm

b

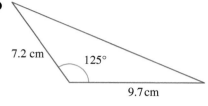

7.2 cm

125°

9.7cm

c

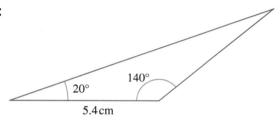

140°

20°

5.4cm

9 *RST* is a triangle.

$RS = 4.2$ cm, $ST = 8.4$ cm and angle $RST = 60°$

a Make an accurate drawing of triangle *RST*.

b Measure side *RT*.

c Measure the other angles.

d What type of triangle is *RST*?

10 Make accurate drawings of these triangles.

In each case you may need to work out an angle before you draw it.

a

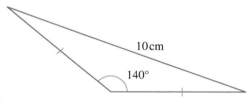

10cm

140°

This triangle is isosceles
so it has two equal angles.

b

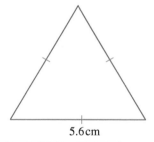

5.6cm

This triangle is equilateral
so it has three equal angles.

11 a Make an accurate drawing of this quadrilateral.

b Measure the length *AD*.

Start by drawing line *BC* and then
measure the two given angles.

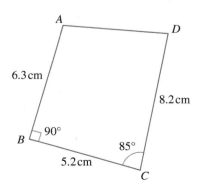

A

D

6.3 cm

8.2cm

B

90°

85°

5.2cm

C

Drawing rectangles and squares

To draw rectangles and squares you use your set square to measure 90° at each corner of the shape.

Worked example 7

Draw a rectangle 7.5 cm by 4.8 cm.

Use your ruler to draw a base line 7.5 cm long.

Position your set square at the end of the line.
Draw a perpendicular to the base line.
Mark a point 4.8 cm up the line.

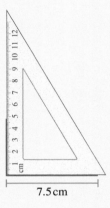

Repeat with the set square at the other end of the base line.

Join the two points to complete the rectangle.

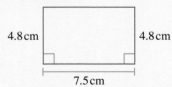

Drawing regular polygons

In Chapter 3 you learned about some polygons.

The sides of a regular polygon are equal and the angles are all equal.

To draw a regular polygon you need the length of each side and the internal angle.

Worked example 8

Draw a regular polygon with sides of length 4 cm long and internal angle of 120°.

What type of polygon have you drawn?

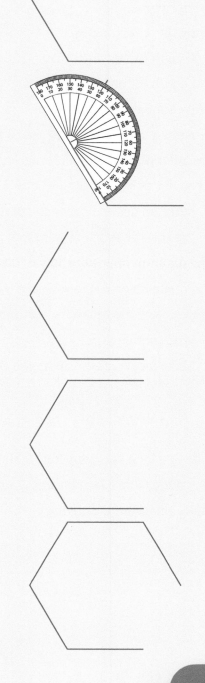

Draw two lines 4 centimetres long at an angle of 120°.

Place the protractor at the end of one of the lines.

Mark an angle of 120°.

Draw another line 4 centimetres long.

Your drawing should now look like this.

Place the protractor at the end of the new line.

Mark another 120° angle.

Draw another 4 cm line.

The drawing should now look like this.

Another 120° angle and 4 cm line will give a drawing like this.

Finally join up the two ends to complete the hexagon.

Check that the final line is 4 cm long.

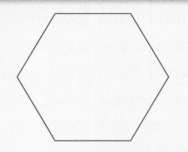

You have drawn a regular hexagon.

Exercise 9.2c

1 Draw these rectangles.

    a Length 6 cm, width 4 cm   **b** Length 7.4 cm, width 2.8 cm

    c Length 3.8 cm, width 1.9 cm

2 Draw a square with 5 cm sides.

3 a Draw a square with 8.4 cm sides.   **b** Write down the length of the diagonal.

4 A regular pentagon has internal angles of 108°. Draw a regular pentagon with side length 5 cm.

5 A regular polygon has sides of 4.5 cm and internal angles of 135°.

    a Make an accurate drawing of the polygon.

    b What is the name of this type of polygon?

6 The diagram opposite shows a design for a badge.

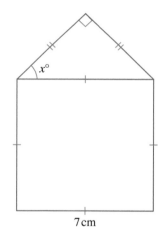

    It is made up of a square with a right-angled isosceles triangle.

    Each side of the square is 7 centimetres long.

    a Work out angle x.

    b Make an accurate drawing of the design.

> The triangle is isosceles so the angle at the right hand side is also $x°$.

7 A design for a company logo is shown opposite.

    It is made from a regular hexagon, two squares and a rectangle.

    Make an accurate drawing of the logo.

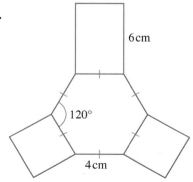

9.3 Solids

Some solids have faces that are polygons. A solid like this is called a polyhedron.

Here are some examples.

A cube has square faces.

It has 6 faces.

It has 8 vertices.

It has 12 edges.

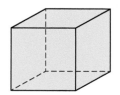

A triangular pyramid has triangular faces.

It has 4 faces.

It has 4 vertices.

It has 6 edges.

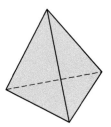

A cuboid has rectangular faces.

It has 6 faces.

It has 8 vertices.

It has 12 edges.

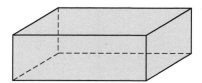

A prism is the same shape along its whole length.

It has a constant cross-section.

A cuboid is a rectangular prism.

If you cut across it parallel to the end face, the shape you find is always a rectangle.

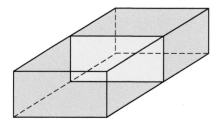

Worked example 1

This is pentagonal prism.
Write down the number of faces, vertices and edges.

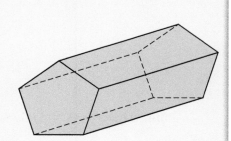

There is a face at each end and 5 faces on the sides.

That makes 7 faces altogether.

There are 5 vertices at each end.

That makes 10 vertices altogether.

There are 5 edges around each end and 5 that run along the prism.

That makes 15 edges altogether.

Exercise 9.3

1 Here are some more solids.

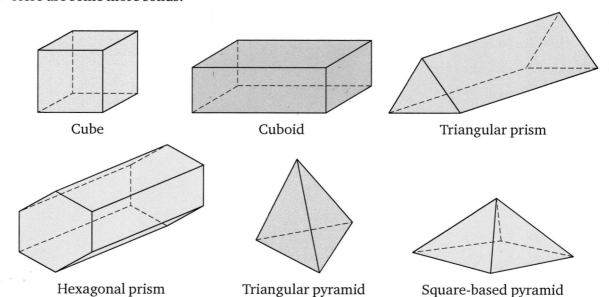

Cube   Cuboid   Triangular prism

Hexagonal prism   Triangular pyramid   Square-based pyramid

Copy and complete this table. The cube has been done for you.

Name	Number of faces	Number of vertices	Number of edges	
Cube	6	8	12	
Cuboid				
Triangular prism				
Hexagonal prism				
Triangular pyramid				
Square-based pyramid				

2 This is an octahedron. It has 8 equilateral triangular faces.

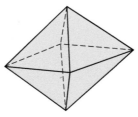

Octahedron

Count the vertices and edges and add this to your table from question **1**.

3 Let f be the number of faces, v the number of vertices and e the number of edges.

For example, for a cube $f = 6$, $v = 8$ and $e = 12$

a For each solid in the table find the value of $f + v - e$

Put the answer in the spare column in the table.

For the cube, $6 + 8 - 12 = 2$, so you put 2 in the final column.

b What do you notice?

Review

1 Find the angles marked with letters.

a

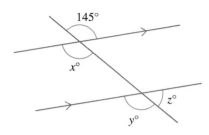

b

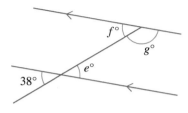

c

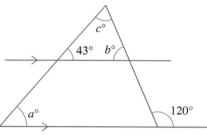

d

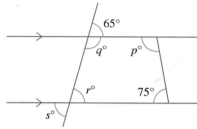

Not drawn accurately

2 *ABC* is a triangle.

Use your ruler and protractor to measure lengths and angles. Copy and complete the following:

a *AB* = _____

b *BC* = _____

c Angle *ABC* = _____

d Angle *ACB* = _____

e Angle *BAC* = _____

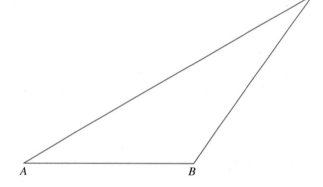

f What do your three angles add up to? Is this what you expected?

3 Make accurate drawings of these triangles.

a

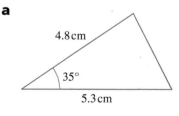

b

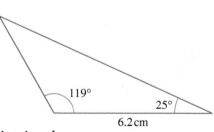

c

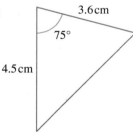

4 Make an accurate drawing of this triangle.

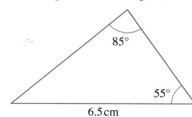

Work out the missing angle first.

5 *ABC* is an isosceles triangle with $AB = AC = 7.2$ cm

Angle $BAC = 70°$

D is the midpoint of *AB*.

DE is parallel to *BC*.

$DE = 6.4$ cm

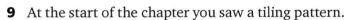

a Make an accurate copy of this diagram.

b Measure the length *AE*.

6 a Draw a rectangle with length 5.9 cm and width 3.4 cm.

b Measure the length of the diagonal.

7 a Draw a square with sides 4.8 cm.

b Measure the length of the diagonal.

8 This design is based on a regular octagon with a square and an equilateral triangle.

The internal angle of a regular octagon is 135°.

All of the sides are 4 cm.

Make an accurate drawing of the design.

9 At the start of the chapter you saw a tiling pattern.

Here is part of a pattern based on regular hexagons and equilateral triangles.

Make an accurate drawing of part of this pattern and colour it in.

10 Write down the correct names of these solids.

a

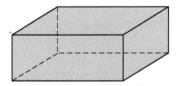

b

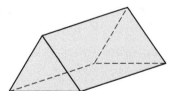

c

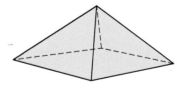

10 Presenting, interpreting and discussing data

Chapter 18 covers collecting data
Chapter 6 covers processing data

Learning outcomes

- Draw and interpret pictograms.
- Draw and interpret bar-line graphs and bar charts.
- Draw and interpret frequency diagrams for grouped discrete data.
- Draw and interpret simple pie charts.
- Draw conclusions based on the shape of graphs.

Misleading graphs

You may have heard the saying 'a picture is worth a thousand words'. So it is with graphs. A simple graph can often save you a lot of words.

However, you may need to look carefully at graphs, as the story may not always be as it first appears. Both these graphs show the same information.

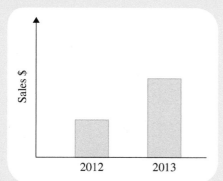
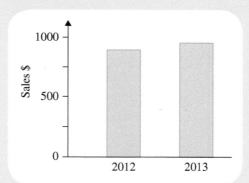

The first graph has no vertical scale, so it is not possible to read any values. It does look as though there is a big increase from 2012 to 2013.

The second graph has a vertical scale. You can see that the increase from 2012 to 2013 is small compared to the height of the rectangles.

10.1 Pictograms and frequency diagrams

Pictograms

A pictogram uses pictures to represent data. Each picture represents a number of items. There is a key, which tells you how many items each picture represents.

Worked example 1

This pictogram shows the sales of some items from a shop during one lunch time.

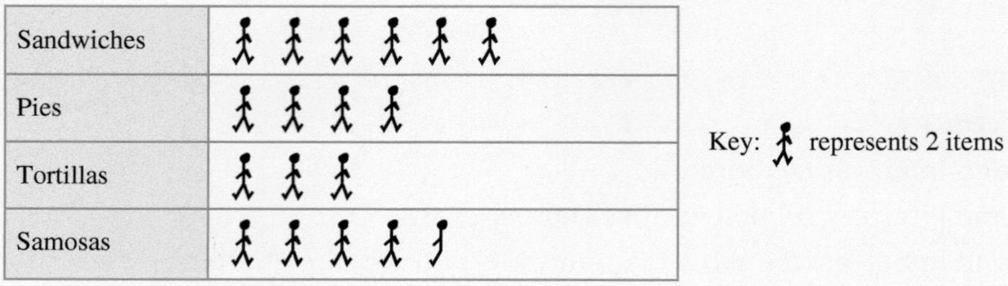

Sandwiches	🚶 🚶 🚶 🚶 🚶 🚶
Pies	🚶 🚶 🚶 🚶
Tortillas	🚶 🚶 🚶
Samosas	🚶 🚶 🚶 🚶 🚶

Key: 🚶 represents 2 items

Find the number of sales for each item.

Sandwiches have 6 figures so 6 × 2 = 12 were sold

Pies have 4 figures so 4 × 2 = 8 were sold

Tortillas have 3 figures so 3 × 2 = 6 were sold

Samosas have $4\frac{1}{2}$ figures so $4\frac{1}{2} \times 2 = 9$ were sold

Worked example 2

The table shows the number of DVDs bought by the students in class 5A last week.

Day	Number bought
Monday	16
Tuesday	18
Wednesday	8
Thursday	11
Friday	25

Show this information in a pictogram. Use the symbol ⬤ to represent 4 DVDs.

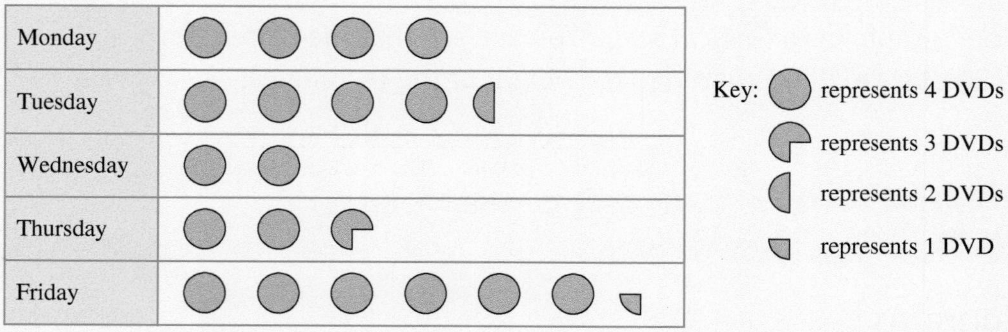

Monday	⬤ ⬤ ⬤ ⬤
Tuesday	⬤ ⬤ ⬤ ⬤ ◖
Wednesday	⬤ ⬤
Thursday	⬤ ⬤ ◕
Friday	⬤ ⬤ ⬤ ⬤ ⬤ ⬤ ◿

Key: ⬤ represents 4 DVDs
◕ represents 3 DVDs
◖ represents 2 DVDs
◿ represents 1 DVD

Bar charts

A bar chart is an example of a frequency diagram.

The bars can be horizontal or vertical.

Bar charts must have:

- a title
- labels on each axis
- values on the frequency axis, evenly spaced starting at zero
- bars of equal width
- equal gaps between the bars
- a label under each bar.

Worked example 3

The frequency table shows information about some of the CDs borrowed from a library.

Type of music	Frequency
Jazz	5
Dance	2
Trance	1
Reggae	12

Show this information on a bar chart.

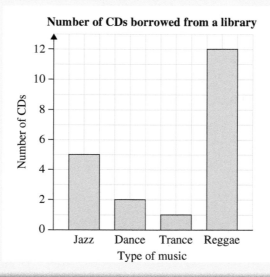

Bar-line graphs

Bar-line graphs are another example of a frequency diagram.

They are very similar to bar charts, but they use thick lines instead of bars.

Worked example 4

Luca counts the number of people in a car as they drive past his school.
His results are shown in the frequency table.

Number of people in a car	Frequency
1	3
2	5
3	4
4	3
5	1

Show this information in a bar-line graph.

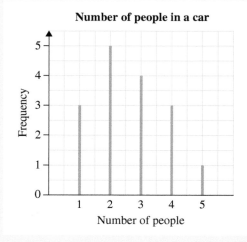

Frequency diagrams for grouped discrete data

Data that covers a large range of values is often grouped. A frequency diagram is very similar to a
bar chart. You must use the same rules as for drawing a bar chart.

If you use a frequency diagram it makes finding the modal class easy. The modal class is the group
with the highest frequency.

Worked example 5

A class of 30 students did a test. The marks are shown in the grouped frequency table.

Marks	Frequency
1–10	3
11–20	6
21–30	10
31–40	7
41–50	4

Draw a frequency diagram to show these marks.

Write down the modal class.

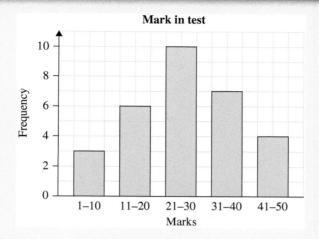

The modal class is 21–30 marks.

Exercise 10.1

1 The frequency table shows the numbers of tickets sold by a ticket office.

Draw a bar chart to show this information.

Type of ticket	Frequency
Child	6
Adult	12
Student	8
Family	5

2 The bar chart shows information about the number of birds visiting a garden one morning.

a How many blackbirds visited the garden?

b How many silvereyes visited the garden?

c Which type of bird visited the garden five times?

d What is the total number of birds visiting the garden?

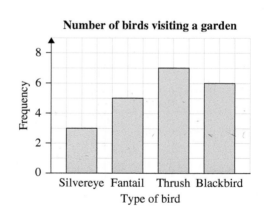

3 Maya visited a duck pond.

She saw 5 grebes, 3 teals, 10 mallards and 6 pochards.

a Put this information into a frequency table.

b Draw a bar chart to show this information.

c Does the table or the bar chart show this information better? Explain your answer.

4 The frequency diagram shows the number of toys on shelves in a shop.

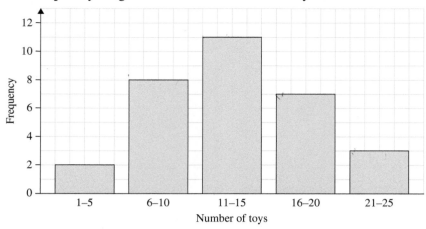

a How many shelves have between 11 and 15 toys?

b How many shelves have fewer than 6 toys?

c How many shelves have more than 15 toys?

d How many shelves are shown in the frequency diagram?

5 Sunil did a survey of the number of people in cars outside his school.

The frequency table shows his results for 10am. The bar-line graph shows his results for 3.30pm.

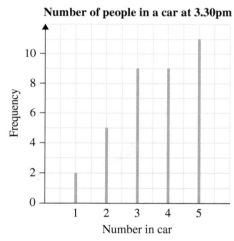

Number of people in a car at 3.30pm

Number of people in a car at 10am	Frequency
1	10
2	3
3	2
4	2
5	1

a Draw a bar-line graph for the 10am results.

b Copy the following sentences and fill in the gaps.

The graph for _____ generally has fewer people in the cars.

The graph for _____ generally has more people in the cars.

The graph for _____ is most likely to be when parents have picked up their children from school.

The graph for _____ could be when people are driving past on their way to do some shopping.

6 The table shows the number of items bought when a group of people go shopping.

Items	1–10	11–20	21–30	31–40	41–50	51–60	61–70	71–80
Frequency	4	7	12	11	10	15	8	3

Draw a frequency diagram to show this information.

7 Mr Patel's class and Mrs Kapoor's class did the same test.

The grouped frequency table shows the marks of Mrs Kapoor's class.

The frequency diagram shows the marks of Mr Patel's class

Mark	Frequency
1–10	2
11–20	5
21–30	9
31–40	6
41–50	5

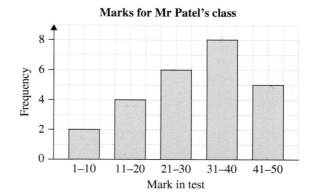

Marks for Mr Patel's class

a Draw a frequency diagram for the marks of Mrs Kapoor's class.

b How many students are in each class?

c Which class appears to have the better marks in the test? Give a reason for your answer.

8 There are four roads in a village.

The pictogram shows the number of houses on three of the roads.

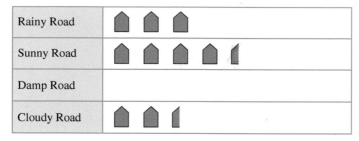

Key: 🏠 represents four houses

a How many houses are there on Cloudy Road?

b How many more houses are there on Sunny Road than Rainy Road?

c Altogether there are 60 houses in the village. Copy and then complete the pictogram.

9 The pictogram shows the amount of money a school has raised for charity.

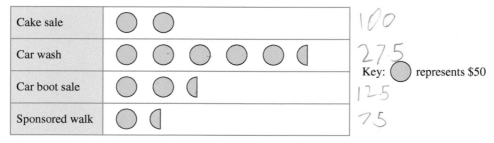

Key: ⬤ represents $50

a Which event raised most money?

b How much money did the car boot sale raise?

c How much more money did the car wash raise than the sponsored walk?

d Altogether the school wants to raise $1100.

One of the teachers is sponsored to run a marathon of 42 kilometres.

He raises the rest of the money.

How much is he sponsored for each kilometre?

> Start by working out the total raised by the cake sale, car wash, car boot sale and sponsored walk. The rest of the $1100 is raised by the marathon runner.

10.2 Pie charts

A **pie chart** shows information on a circle. The circle is divided into sectors. The angles must add up to 360°. You must label each sector and draw the angles accurately.

Worked example 1

The frequency table shows how 18 students travel to school each day.

Draw a pie chart to represent this information.

Method of transport	Frequency
Walk	8
Car	2
Bus	5
Train	1
Other	2

The total number of students is 18.

The 360° at the centre of the circle must be divided between the 18 students.

$360° \div 18 = 20°$

Each student will be represented by 20°.

You multiply each frequency by 20° to find the angle for each sector.

Method of transport	Frequency	Angle
Walk	8	$8 \times 20° = 160°$
Car	2	$2 \times 20° = 40°$
Bus	5	$5 \times 20° = 100°$
Train	1	$1 \times 20° = 20°$
Other	2	$2 \times 20° = 40°$

> Before you draw the pie chart always check the angles add up to 360°
> ($160° + 40° + 100° + 20° + 40° = 360°$).

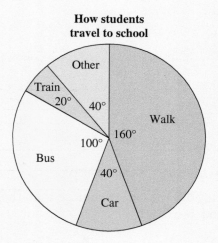

How students travel to school

Worked example 2

The pie chart shows the number of food items sold by a café one lunchtime.

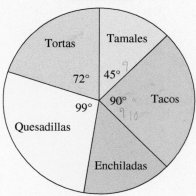

**Food items sold by a café
one lunchtime**

The café sold five tamales.

a What is the angle for enchiladas?

b How many tacos were sold?

c How many food items were sold altogether?

a The angles at the centre of the circle add up to 360°.

The angle for enchiladas = 360° − (45° + 90° + 99° + 72°) = 360° − 306° = 54°

b Five tamales are represented by 45°. So, each tamale is represented by 45° ÷ 5 = 9°

The sector for tacos has an angle of 90°.

90° ÷ 9° = 10 tacos.

c The angle for all the food items is 360°. 360° ÷ 9° = 40 food items.

Exercise 10.2

1 A café sells ice cream in four flavours.

The frequency table shows the number sold of each flavour one morning.

Flavour	Frequency
Vanilla	20
Strawberry	10
Toffee ripple	2
Banana	4

Show this information on a pie chart.

2 The pie chart shows information about how 72 people travel to work.

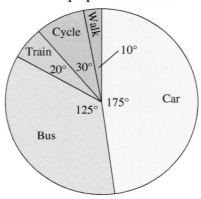

How 72 people travel to work

a What angle represents 1 person?

b Copy and complete the frequency table.

Method of transport	Frequency
Car	
Bus	
Train	
Cycle	
Walk	

3 The pie chart shows the favourite lesson of a class of students.

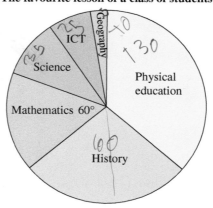

The favourite lesson of a class of students

Five students chose mathematics as their favourite lesson.
How many students are in the class?

4 A class of 20 children are asked about their pets. None of the children have more than one pet.

5 have a cat, 6 have a hamster, 4 have a bird, 2 have a gerbil.

The rest have no pet.

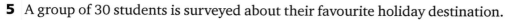

a How many children have no pet?

b Draw a pie chart to show this information

5 A group of 30 students is surveyed about their favourite holiday destination.

The results are shown in the pie chart.

The parents of these students are also surveyed.

The results from the parents are shown in the bar chart.

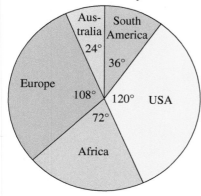

Students' favourite holiday destinations

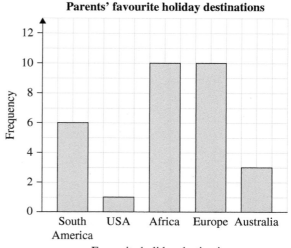

Compare the results of the two surveys.

> You need to show the results of the two surveys in the same way before you can compare them.

6 A group of students is asked about their favourite drink. The table shows the results.

Drink	Frequency	Angle for pie chart sector
Fizzy drinks	18	
Squash	24	
Fruit juice		60
Milk		
Total	**60**	

Find the missing numbers in the table.

Draw the pie chart.

Review

1 A class of students is asked about their favourite sport.

The results are shown in the bar chart.

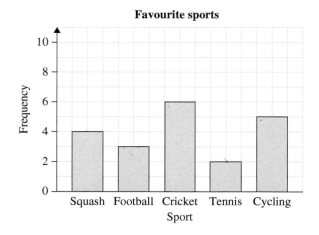

a How many students chose football as their favourite sport?

b How many more students chose cycling than tennis?

c Which is the most popular sport?

d How many students are in the class?

2 The graph shows the number of letters in words in the first few sentences of a book.

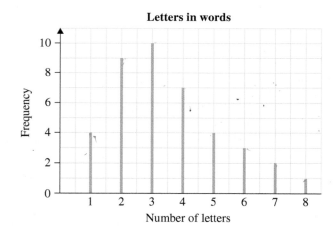

a What is the name given to this type of graph?

b How many words have seven letters?

c What is the most common number of letters in these words?

d How many letters are in the longest word?

e How many words have fewer than 3 letters?

f How many words are represented in the graph?

3 The table shows the number of different types of sandwiches sold by a café at lunchtime.

Sandwich	Frequency
Egg	20
Hummus	18
Tuna	11
Chicken	9
Cheese	13

Draw a pictogram to show this information.

Use one symbol to represent four sandwiches.

Choose your own symbol.

4 A teacher recorded the number of days' absence by her students for one term.

The results are shown in the grouped frequency table.

Days' absence	Number of students
0–4	8
5–9	7
10–14	5
15–19	3
20–24	2

Draw a frequency diagram to show this information.

5 A group of 20 students are asked what type of puzzles they like to solve.

The results are shown in the pie chart.

a Show that four students like crosswords.

b Copy and complete this sentence.

25% of students like to do _____ puzzles.

c How many students do not like to do puzzles?

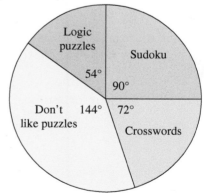

6 The frequency table shows the number of cars owned by people in a small village.

Number of cars owned	Frequency
0	4
1	9
2	8
3	2
4	1

Draw a pie chart to show this information.

7 Each of the following is a bar chart with at least one problem (they may have more than one). For each bar chart, write down all the problems you can see.

a

b

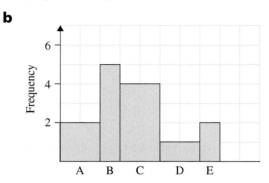

c

d

Number of students getting grades in a test

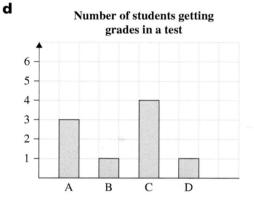

8 A company uses this bar chart to show its profits.

a The bar chart is misleading. What is wrong with the bar chart?

b Draw an improved bar chart to show the same information.

Profits soar

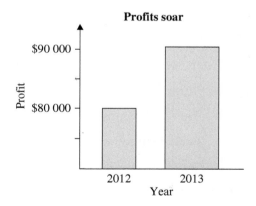

11 Area, perimeter and volume

Painting and decorating

Before painting the walls in a room you have to work out how much paint to buy.

On the paint tins it tells you how much paint you need for a given area.

If you are planning to use wallpaper you need to find the perimeter of the room.

This chapter tells you all about area and perimeter.

11.1 Area and perimeter

Area is a measure of how big a surface is. These three shapes look very different but they all have the same area as they cover the same amount of space on the page.

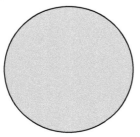

Units of area

A square that is 1 cm by 1 cm has an area of 1 square centimetre.

You write this as $1\,cm^2$.

You would use cm^2 for measuring areas such as the page of this book.

1 cm
1 cm

A square that is 1 mm by 1 mm has an area of 1 square millimetre.

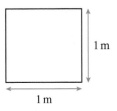

Not drawn to scale

You write this as 1 mm².

You would use mm² for measuring small areas such as the area of a postage stamp.

A square that is 1 m by 1 m has an area of 1 square metre.

You write this as 1 m².

It is too big to draw full size in this book.

You would use m² for measuring large areas such as a football pitch.

Not drawn to scale

Area of rectangles

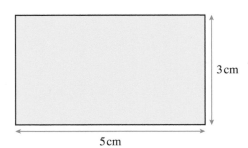

This rectangle is 5 centimetres long by 3 centimetres wide.

This diagram shows that you can fit 15 square centimetres inside it. The area is 15 cm².

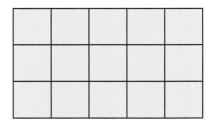

You can see that there are 3 rows of 5 squares.

You can use multiplication to find the area of the rectangle.

$$5 \times 3 = 15$$

The area is 15 cm².

This rectangle is 4 cm by 2 cm.

$$4 \times 2 = 8$$

The area is 8 cm².

You can see on the diagram that this is correct.

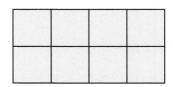

The length of this rectangle is l.

The width is w.

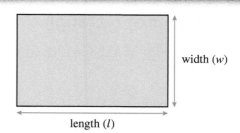

width (w)

length (l)

The area is length × width

The formula for the area, A, of a rectangle is:

$$A = l \times w \quad \text{or} \quad A = lw$$

> Remember that the units must match.
> If the lengths are in cm, the area is in cm².
> If the lengths are in mm, the area is in mm².
> If the lengths are in m, the area is in m².

Worked example 1

Find the area of these rectangles. They are not drawn to scale.

a

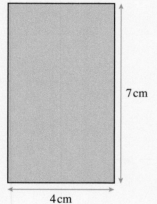

7 cm

4 cm

b

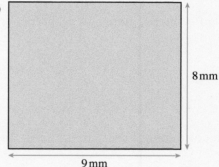

8 mm

9 mm

a Area = $l \times w$
Area = 4×7
Area = $28\,\text{cm}^2$

b Area = $l \times w$
Area = 9×8
Area = $72\,\text{mm}^2$

> Note that the units are millimetres so the answer is in mm².

Worked example 2

A rectangle is 9 cm long. The area is 54 cm².

How wide is the rectangle?

$l \times w = A$ ⟵ This is the formula for the area of a rectangle.

$9 \times w = 54$

$w = 54 \div 9$

$w = 6$

The width is 6 cm. ⟵ Don't forget to write in the units.

Perimeter of rectangles

The **perimeter** of a shape is the distance around the outside.

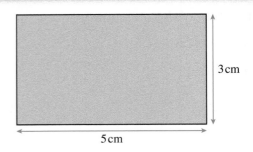

For this rectangle the distance around the outside is 5 cm + 3 cm + 5 cm + 3 cm

The perimeter of the shape is 16 cm.

You add together two lengths and two widths.

Note that the perimeter is a length not an area, so the units are mm, cm or m (not mm², cm² or m²).

The length of this rectangle is l.

The width is w.

The perimeter is 2 × length + 2 × width

The formula for the perimeter, P, of a rectangle is:

$$P = 2l + 2w$$

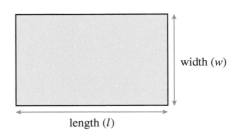

Worked example 3

Find the perimeter of these rectangles. They are not drawn to scale.

a
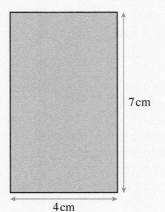

b
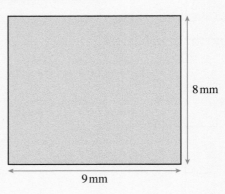

a Perimeter = $2l + 2w$

Perimeter = $2 \times 4 + 2 \times 7$

Perimeter = 22 cm

b Perimeter = $2l + 2w$

Perimeter = $2 \times 9 + 2 \times 8$

Perimeter = 34 mm

Note that the units are millimetres.

Converting units

This is a 1 centimetre square.

The area is $1 \times 1 = 1\,\text{cm}^2$

This is the same square measured in millimetres.

You saw in Chapter 7 that $1\,\text{cm} = 10\,\text{mm}$

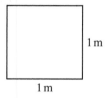

The area of this square is $10 \times 10 = 100\,\text{mm}^2$

You can see that $\mathbf{1\,cm^2 = 100\,mm^2}$

You can do the same with a bigger square.

These two squares are the same size. The left hand one is measured in metres, the right hand one is measured in centimetres.

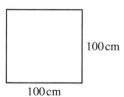

Not drawn to scale

Area $= 1 \times 1 = 1\,\text{m}^2$ Area $= 100 \times 100 = 10\,000\,\text{cm}^2$

You can see that $\mathbf{1\,m^2 = 10\,000\,cm^2}$

Exercise 11.1

The diagrams in this exercise are not drawn to scale.

1 Find the area of each of these rectangles. State the units in your answers.

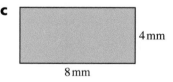

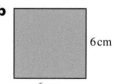

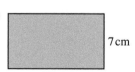

2 Find the perimeter of each of the rectangles in question 1.

3 A rectangular wall is 3.2 metres long and 2.4 metres high.

    a What is the area of the wall?

    b What is the perimeter of the wall?

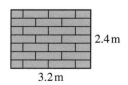

4 A farmer has a rectangular field 36 metres long and 17 metres wide.

What length of fence is needed to go around the edge of the field?

5 A small tin of paint covers an area of 10 m².

Is the tin big enough for one coat of paint on a ceiling 3.4 metres by 2.8 metres?

6 Toya has a piece of material 2 metres by 2 metres.

Fatima has a piece of material 5 metres by 75 centimetres.

Which of the girls has the piece with the bigger area?

> Make sure the units are all the same before calculating the area.

7 This rectangle is 12 cm long.

The area is 60 cm².

How wide is the rectangle?

60 cm²

12 cm

8 The area of this square is 81 mm².

How long is each side?

81 mm²

9 Use these diagrams to find how many square millimetres there are in 1 square metre.

1 m

1 m

1000 mm

1000 mm

10 This Indian postage stamp measures 22 mm by 34 mm.

   a Find the area in mm².

   b Find the area in cm².

   c Find the perimeter in mm.

   d Find the perimeter in cm.

11 a This rectangle has an area of 24 cm².

   Find another rectangle with the same area.

   b Find some more rectangles with an area of 24 cm².

4 cm

6 cm

   c How many such rectangles are there if the sides are all whole numbers?

> Think of all the factor pairs of 24.

   d How many are there if the sides can be fractions of a square?

12 The rectangle in question **11** has a perimeter of 20 cm.

   a Investigate other rectangles with the same perimeter.

   Only consider rectangles whose sides are whole numbers.

   b Which of the rectangles has the largest area?

> Be systematic. Draw a rectangle 1 cm wide, then draw one 2 cm wide, then 3 cm, etc.

13 Work out how many square metres there are in one square kilometre.

It may help to draw diagrams like those in question **9**.

11.2 Compound shapes

A compound shape can be made out of two or more rectangles.

Here is an example of a compound shape.

Area of compound shapes

This is how to find the area of a compound shape.

This shape is made out of rectangles.

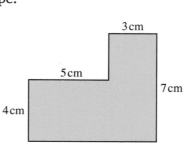

Draw lines across the shape to divide it into rectangles.

You can do this in several different ways as shown here.

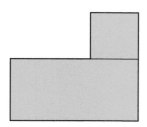 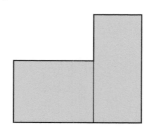 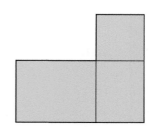

It does not matter which arrangement you choose.

Using the first diagram:

It is divided into two rectangles.

Each rectangle is labelled with a capital letter.

You work out each area separately.

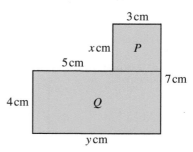

Rectangle P

The side of rectangle Q is 4 cm. The length $x = 7 - 4 = 3$

Rectangle P is 3 cm across. The area of $P = 3 \times 3 = 9$ cm²

Rectangle Q

The length $y = 5 + 3 = 8$ because $5 + 3 = y$, the total distance across the shape

The area of $Q = 4 \times 8 = 32$ cm²

Total area $= 9 + 32 = 41$ cm²   you add the two areas to get the total

Perimeter of compound shapes

To find the perimeter you work out every side and add them all together.

You have already seen that the side of *P* is 3 cm and the base of *Q* is 8 cm.

Here is the shape with all of the measurements shown.

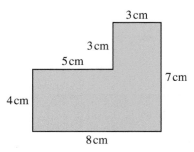

To find the perimeter you add together all of the sides.

Start at a corner and add the sides in turn as you move round the shape.

For example, moving anticlockwise from the bottom left hand corner gives:

$$8 + 7 + 3 + 3 + 5 + 4 = 30$$

There are six sides so check you are adding six numbers.

Perimeter = 30 cm

Worked example 1

a Find the area of this shape.

b Find the perimeter.

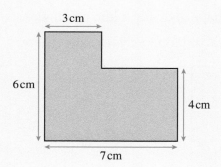

a Divide the shape into two rectangles, *X* and *Y*.

The area marked *X* is 6 cm by 3 cm.

Area $X = 3 \times 6 = 18$ cm²

$w = 7 - 3 = 4$

Area $Y = 4 \times 4 = 16$ cm²

Total area $= 18 + 16 = 34$ cm²

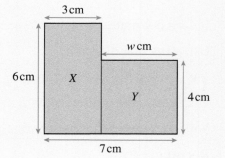

b $w = 4$ from part **a**

$x = 6 - 4 = 2$

Perimeter $= 7 + 4 + 4 + 2 + 3 + 6 = 26$ cm

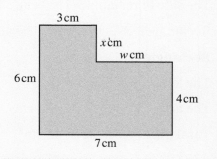

Sometimes you can find the area by subtraction.

The next example shows you how.

Worked example 2

a Find the area of this shape.

b Find the perimeter.

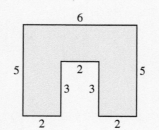

All units are in centimetres

a You can think of this shape as a large rectangle with a small rectangle cut out.

Area of the large rectangle is $6 \times 5 = 30$

Area of the small rectangle is $2 \times 3 = 6$

Area of the shaded shape is $30 - 6 = 24\,cm^2$

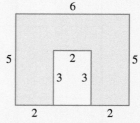

All units are in centimetres

> The small rectangle has been cut out so you **subtract** from the large rectangle to find the area.

b You find the perimeter by adding all the sides.

$6 + 5 + 2 + 3 + 2 + 3 + 2 + 5 = 28\,cm$

> Check – there are eight sides so there are eight numbers added.

Exercise 11.2

The diagrams in this exercise are not drawn to scale.

1 Find the area of each of these shapes. State the units in your answers.

a

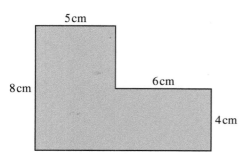

b

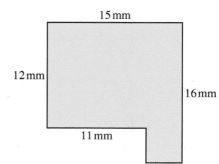

c

10cm

3cm

4cm

7cm

d

12cm

3cm

12cm

2cm

e

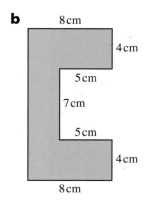

8cm

2cm

5cm

5cm

4cm

3cm

7cm

2 Find the perimeter of each of the shapes in question **1**.

3 Find the area of each of these shapes.

a

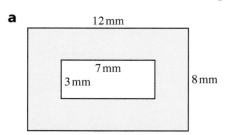

12mm

7mm

3mm

8mm

b

8cm

4cm

5cm

7cm

5cm

4cm

8cm

4 This diagram shows a garden. It is a large lawn with a square pond in it.

   a Find the total area of the garden.

   b Find the area of the pond.

   c Find the area of the lawn.

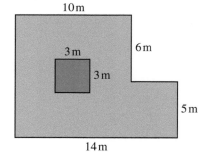

10m

3m

6m

3m

5m

14m

5 This is a plan of part of a house.

   a Find the area of the kitchen.

   b Find the area of the hall.

   c Find the area of the living room.

      Subtract the areas of the kitchen and the hall from the total area.

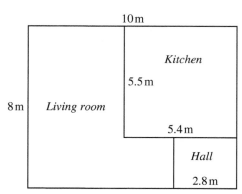

10m

Kitchen

5.5m

8m   Living room

5.4m

Hall

2.8m

6 Use squared paper for this question.

The shape shown here has an area of 5 cm².

Using only whole squares, investigate other shapes with an area of 5 cm².

  a How many shapes can you find?

  b Work out the perimeter of each shape that you find.

> The squares must meet edge to edge, not corner to corner.

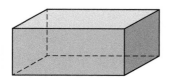 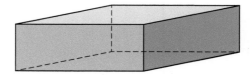

7 A machine stamps small brackets out of a sheet of metal.

This diagram shows one bracket.

  a Work out the area of one bracket.

The machine uses a sheet of metal 1 metre by 1 metre.

It cuts 400 brackets out of one sheet.

  b Work out the area of 400 brackets.

  c What area of metal is wasted? Give the answer in cm².

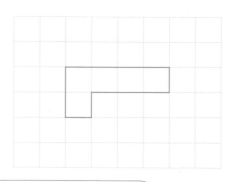

8 cm
2 cm
6 cm
3 cm

Not drawn to scale

> Remember that
> 1 m² = 10 000 cm².

11.3 Cuboids

A **cuboid** is a three-dimensional shape with six rectangular faces.

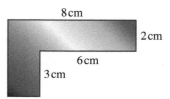

A **cube** is a special cuboid. It has six square faces.

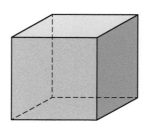

Volume

Volume is a measure of how much space a three-dimensional object takes up. It is measured using cubic millimetres, cubic centimetres or cubic metres.

A cube that is 1 mm by 1 mm by 1 mm has a volume of 1 cubic millimetre.

1 mm
1 mm 1 mm
Not drawn to scale

You write this as $1\,mm^3$.

You use mm^3 for very small volumes.

A cube that is 1 cm by 1 cm by 1 cm has a volume of 1 cubic centimetre.

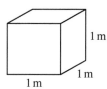
1 cm
1 cm
1 cm

You write this as $1\,cm^3$.

cm^3 is the most common measure of volume.

A cube that is 1 m by 1 m by 1 m has a volume of 1 cubic metre.

You write this as $1\,m^3$.

You use m^3 for very large volumes.

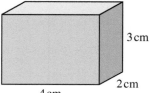
1 m
1 m
1 m
Not drawn to scale

Finding volumes of cuboids

This cuboid is 4 cm wide by 2 cm wide by 3 cm high.

To measure the volume you need to find how many centimetre cubes would fill the cuboid.

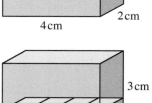
3 cm
4 cm
2 cm

This diagram shows one layer of centimetre cubes in the cuboid.

There are $4 \times 2 = 8$ cubes in this bottom layer.

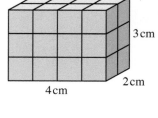
3 cm
4 cm
2 cm

The next diagram shows the cuboid filled with centimetre cubes.

There are 3 layers of cubes. Each layer has 8 cubes.

There are $8 \times 3 = 24$ cubes altogether.

The volume is $24\,cm^3$.

You can find the volume by multiplying $4 \times 2 \times 3 = 24$
The volume is $24\,cm^3$.

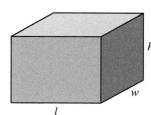
3 cm
4 cm
2 cm

The length of this cuboid is l. The width is w. The height is h.

The volume is length × width × height.

The formula for the volume of a cuboid is:

$$V = l \times w \times h \qquad \text{or} \qquad V = lwh$$

Worked example

Find the volume of a cube with edges 4 cm long.

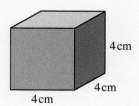

4 cm

4 cm

4 cm

The formula for the volume is $V = lwh$

For this cuboid $l = w = h = 4$

$V = 4 \times 4 \times 4 = 64$

The volume is 64 cm³.

Don't forget to show the units. If the lengths are in cm, the volume is in cm³.

Exercise 11.3

1 Find the volume of each of these cuboids.

a

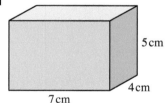

5 cm

4 cm

7 cm

b

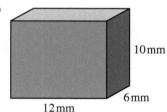

10 mm

6 mm

12 mm

c

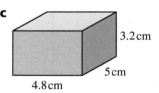

3.2 cm

5 cm

4.8 cm

2 Find the volumes of these cubes.

a

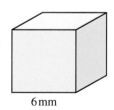

6 mm

b

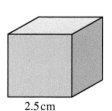

2.5 cm

3 A box file is 36 cm by 25 cm by 8 cm.

Find the volume of the file.

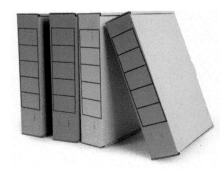

4 A water tank is 40 centimetres long, 18 centimetres wide and 25 centimetres deep.

   a Calculate the volume of the tank.

   b 1 cm³ of water has a mass of 1 gram.

   What is the mass of the water in the tank?

   Give the answer in kilograms.

> Remember that 1 kg = 1 000 g

5 A cuboid flower pot has a 40 cm square base.

   A gardener tips 0.1 m³ of compost into the pot.

   How deep will the compost be?

> 40 cm = 0.4 m. If the depth is x m then the volume is $0.4 \times 0.4 \times x$ m³

6 These cuboids all have the same volume.

   Work out the lengths of the sides marked with letters.

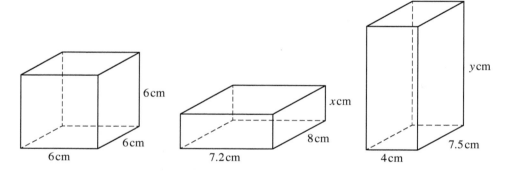

Not drawn to scale

> Work out the volume of the cube then make equations to find the missing heights.

11.4 Surface area

Every cuboid has six rectangular or square faces.

To find the surface area of the cuboid you need to add together the area of each face.

You can use the net of the cuboid to help you.

This cuboid is 4 cm by 2 cm by 3 cm.

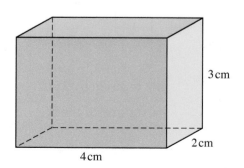

This is the net of the cuboid. It can be used to make the cuboid out of card or paper.

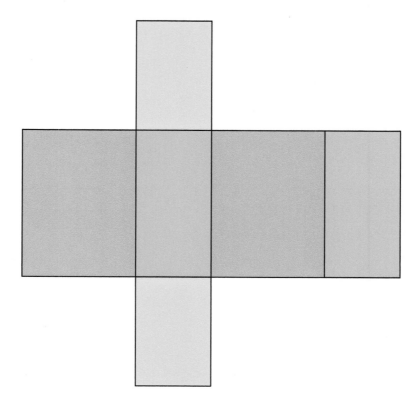

Each pink face is 4 cm by 3 cm. The area of each pink face is $4 \times 3 = 12\,\text{cm}^2$

Each blue face is 4 cm by 2 cm. The area of each blue face is $4 \times 2 = 8\,\text{cm}^2$

Each yellow face is 3 cm by 2 cm. The area of each yellow face is $3 \times 2 = 6\,\text{cm}^2$

The total surface area is:

$12 + 12 + 8 + 8 + 6 + 6 = 52\,\text{cm}^2$

It is easier to add one of each face and then multiply by 2.

Total surface area is $(12 + 8 + 6) \times 2 = 52\,\text{cm}^2$

Worked example

Find the surface area of this cuboid.

Area of the top = $5 \times 5 = 25\,\text{mm}^2$

Area of the front = $5 \times 3 = 15\,\text{mm}^2$

Area of the side = $5 \times 3 = 15\,\text{mm}^2$

Total surface area = $(25 + 15 + 15) \times 2 = 110\,\text{mm}^2$

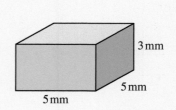

Exercise 11.4

1 Work out the surface area of these cuboids.

a
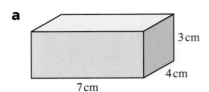
3 cm
4 cm
7 cm

b
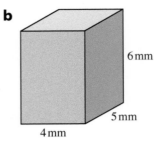
6 mm
5 mm
4 mm

c
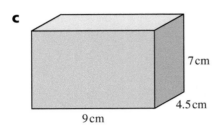
7 cm
4.5 cm
9 cm

d
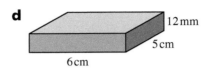
12 mm
5 cm
6 cm

e
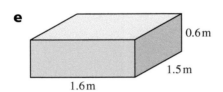
0.6 m
1.5 m
1.6 m

Draw the net of the cuboid first so that you can see the size of each face.

2 Work out the surface area of these cubes.

a
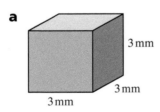
3 mm
3 mm
3 mm

b
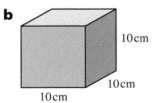
10 cm
10 cm
10 cm

3 Which of these two shapes has the larger surface area? Show all of your working.

a
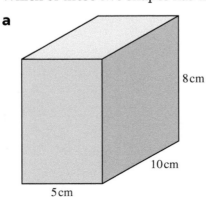
8 cm
10 cm
5 cm

b
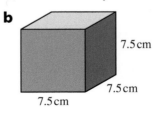
7.5 cm
7.5 cm
7.5 cm

4 The cuboid shown here is measured in millimetres.

    a Work out the surface area in mm².

    b Work out the surface area in cm².

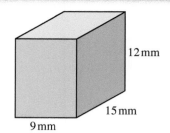

12mm
15mm
9mm

5 A metal storage tank is 2.4 metres long, 1.6 metres wide and 1.2 metres high.

    Work out the total surface area of the tank.

> Drawing a sketch of the tank may help.

6 This solid is a cuboid and a cube glued together.

    The cuboid measures 4 cm by 5 cm by 6 cm.

    The cube has sides of length 2 cm.

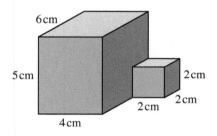

6cm
5cm
4cm
2cm
2cm
2cm

    a Work out the total volume.

    b Work out the total surface area.

> First work out the surface area of each solid and then subtract the parts that are glued together.

7 These cuboids all have the same surface area.

    Work out the lengths of the sides marked with letters.

> Work out the surface area of the cube and then make equations to find the missing heights.

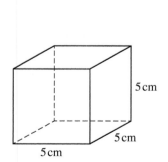

5cm
5cm
5cm

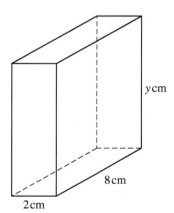

y cm
8cm
2cm

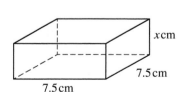

x cm
7.5cm
7.5cm

Review

1 Work out the area of these shapes.

a
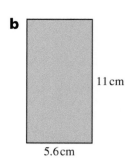
4 mm
7 mm

b
11 cm
5.6 cm

c
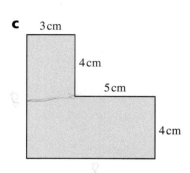
3 cm
4 cm
5 cm
4 cm

d
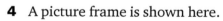
16 mm
4 mm
7 mm
10 mm

e
10 m
6 m
4 m   4 m
3 m   3 m

2 Find the perimeter of each of the shapes in question **1**.

3 This rectangle is 1.8 cm long and 5 mm wide.

   a Work out the area in mm².

   b Work out the area in cm².

   c How many mm² are there in 1 cm²?

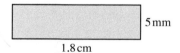
5 mm
1.8 cm

4 A picture frame is shown here.

The outside of the frame is a rectangle 20 cm by 15 cm.

The size of the picture is 14 cm by 9 cm.

Work out the area of the frame, shaded brown in the diagram.

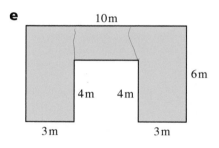
9 cm
14 cm
15 cm
20 cm

5 Work out the volume of each of these cuboids.

a
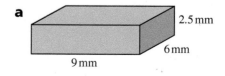
2.5 mm
6 mm
9 mm

b
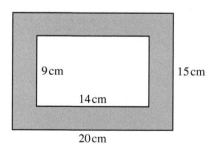
5 cm
4 cm
7 cm

6 Find the surface area of each of the solids in question **5**.

12 Formulae

Using formulae in real life

Formulae are used to solve many real life problems.

If an astronaut wanted to know the time, t, which it would take to travel to the planet Venus he would use the formula:

$$t = \frac{d}{s}$$

where d = the distance between the Earth and Venus and s = the speed of the rocket.

12.1 Deriving formulae

You have already learned about expressions and equations.

$2x + 3y$ is called an **expression**. Note: An expression does not contain an equal sign.

$5x + 1 = 16$ is called an **equation**. (You can solve an equation to find the unknown variable.)

$P = 2l + 2w$ is called a **formula**.

A formula is a rule that connects two or more variables.

The formula for the perimeter of a rectangle can be written in words as:

Perimeter $= 2 \times l$ength $+ 2 \times w$idth

but it is much quicker to write the formula algebraically as:

$P = 2l + 2w$

To **derive** a formula means writing your own formula to connect some variables.

Worked example 1

Jon is 23 years older than Paul.

Jon is J years old and Paul is P years old.

Write a formula connecting J and P.

First write the rule using words: Jon's age = Paul's age + 23

Now write the rule using algebra: $J = P + 23$

Worked example 2

A book costs $7 and a magazine costs $4.

Write down a formula for the total cost C, in dollars, of b books and m magazines.

First write the rule using words: Total cost = cost of books + cost of magazines

    Cost of books = $7 \times b = 7b$

    Cost of magazines = $4 \times m = 4m$

Now write the rule using algebra: $C = 7b + 4m$

Exercise 12.1

1 Cinema tickets cost $7 each.

  a Find the cost of buying 4 tickets.

  b Find the cost of buying 9 tickets.

  c Write down a formula for the total cost C, in dollars, of buying n cinema tickets.

2 A taxi can carry 6 passengers.

  a How many passengers can be carried in 3 taxis?

  b How many passengers can be carried in 7 taxis?

  c Write down a formula for the total number of passengers, P, which can be carried by x taxis.

3 a How many hours are there in 2 days?

  b How many hours are there in 5 days?

  c Write down a formula for the number of hours, N, in x days.

4 A swimming pool contains x litres of water.

  Kiran removes y litres of water from the swimming pool.

  Write down a formula for the number of litres of water, N, remaining in the pool.

5 The instructions for calculating the cooking time, T minutes, for a chicken of mass m kilograms are shown below.

> Allow 40 minutes per kg plus 25 minutes.

Write down a formula connecting T and m.

6

Write down a formula for the total cost C, in dollars, of x ice creams and y ice lollies.

7

Haroon buys x rulers, y pens and z pencils.

Write down a formula for the total cost, C cents.

8

Write down a formula for the total cost T, in dollars, of a adult tickets and c child tickets.

9 A small bus can carry 12 passengers. A large bus can carry 35 passengers.

Write down a formula for the total number of passengers, P, which can be carried by x small buses and y large buses.

10

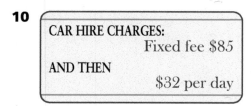

Write down a formula for the total cost, C dollars, of hiring a car for n days.

11 Write down a formula for the perimeter, P, of this six-sided shape.

Explain your answer.

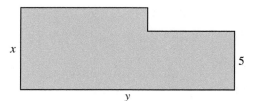

12.2 Substitution into formulae

Substitution into a formula means replacing the letters in a formula by the given numbers.

Worked example 1

$P = 2x + 3y$

Find the value of P when $x = 6$ and $y = 5$

$P = 2x + 3y$ replace the letters with the given numbers

$P = (2 \times 6) + (3 \times 5)$

$P = 12 + 15$

$P = 27$

Worked example 2

$v = u + at$

Find the value of v when $u = 10$, $a = 4$ and $t = 5$

$v = u + at$ replace the letters with the given numbers

$v = 10 + 4 \times 5$ remember to do the multiplication before the addition (BIDMAS)

$v = 10 + 20$

$v = 30$

Exercise 12.2

1 $y = 3x + 2$ Find the value of y when:

    a $x = 6$ **b** $x = 23$ **c** $x = 1.2$ **d** $x = \frac{1}{3}$

2 $f = 2h - 7$ Find the value of f when:

    a $h = 5$ **b** $h = 17$ **c** $h = 3.5$ **d** $h = 2.5$

3 $a = \frac{b}{2} + 5$ Find the value of a when:

    a $b = 8$ **b** $b = 20$ **c** $b = 7$ **d** $b = 4.8$

4 $y = \frac{3x + 5}{2}$ Find the value of y when:

    a $x = 3$ **b** $x = 11$ **c** $x = 2$ **d** $x = 10$

5 $C = 4(d + 1)$ Find the value of C when:

    a $d = 6$ **b** $d = 14$ **c** $d = 0.5$ **d** $d = 1.5$

6 $y = 100 - 3x$ Find the value of y when:

    a $x = 8$ **b** $x = 15$ **c** $x = 24$ **d** $x = 1\frac{1}{3}$

7 $f = 5gh$ Find the value of f when $g = 3$ and $h = 4$

8 $a = bc + d$ Find the value of a when $b = 4$, $c = 7$ and $d = 8$

9 The perimeter, P, of this shape can be found using the formula:

    $P = 2a + 4b$

    Find the value of P when $a = 14$ and $b = 7$

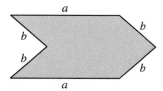

10 **a** Show that the total surface area, A, of the cuboid can be found using the formula:

    $A = 2xy + 2xz + 2yz$

  b Find the value of A when $x = 5$, $y = 3$ and $z = 2$

    A cuboid has six rectangular faces.

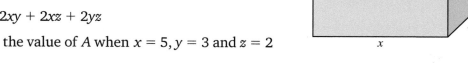

12.3 Further substitution into formulae

Sometimes when you substitute into a formula you obtain an equation to solve.

To solve the equation you need to remember your rules from Chapter 8.

Worked example

$W = 2a + b$

Find the value of a when $W = 17$ and $b = 3$

$W = 2a + b$ replace the letters with the given numbers

$17 = 2a + 3$ subtract 3 from both sides

$14 = 2a$ divide both sides by 2

$a = 7$

Exercise 12.3

1 $y = 5x - 3$

Find the value of x when $y = 17$

2 $f = gh$

Find the value of h when $f = 45$ and $g = 3$

3 $P = 2x + 3y$

Find the value of x when $P = 24$ and $y = 3$

4 $C = 20 + 5x - 2y$

Find the value of x when $C = 40$ and $y = 5$

5 $f = 3gh$

Find the value of h when $f = 30$ and $g = 2$

6 $p = mn - 8$

Find the value of n when $p = 25$ and $m = 3$

7 The cost C, in $, of hiring a boat for n hours is given by the formula:

$$C = 7n + 20$$

Find the value of n if $C = 41$

8 Khadeeja wants to hire a bike.

She can hire a bike from Company A or Company B.

The table shows the cost C, in $, of hiring the bike for n hours from each company.

Company A	$C = 2n + 6$
Company B	$C = 3n + 2$

a Find the cost of hiring a bike for 2 hours from:

i Company A   **ii** Company B.

b Find the cost of hiring a bike for 5 hours from:

i Company A   **ii** Company B.

c Find the value of n for which the cost of hiring the bike from each company is the same.

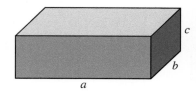

9 a Write down a formula for the total surface area, A, of the cuboid.

b Find the value of b when $A = 158$, $a = 5$ and $c = 3$

Review

1 Sundeep is 9 years older than Raju.

   Sundeep is S years old and Raju is R years old.

   Write a formula connecting S and R.

2 A scientific calculator costs $15 and a graphical calculator costs $36.

   Write down a formula for the total cost C, in dollars, of s scientific calculators and g graphical calculators.

3
> **BOAT HIRE CHARGES:**
>   Fixed fee: $28
>
> **AND THEN**
> $15 per hour

   Write down a formula for the total cost, C dollars, of hiring a boat for n hours.

4 $y = 5x + 4$

   Find the value of y when $x = 8$

5 $y = 2(3x - 7)$

   Find the value of y when $x = 15$

6 $y = \frac{2x}{5} - 8$

   Find the value of y when $x = 15$

7 $p = 3qr$

   Find the value of p when $q = 6$ and $r = 4$

8 $C = 5x + 4y$

   Find the value of x when $C = 38$ and $y = 2$

9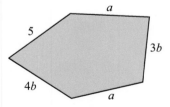

   Write a formula for the perimeter P of the shape.

10 In two years' time, Lesley will be twice the age of Helen.

   Lesley is L years old and Helen is H years old.

   Write a formula connecting L and H.

13 Position and movement

Learning outcomes

- Transform 2-D points and shapes by:
  - reflection in a given line
  - rotation about a given point
  - translation.
- Know that shapes remain congruent after these transformations.
- Read and plot coordinates of points determined by geometric information in all four quadrants.

On reflection

Photographers and artists often use reflection in their work.

This picture shows the Taj Mahal reflected in the Yamuna River at sunset.

The photographer has used the way that water acts as a mirror to make an attractive picture.

13.1 Reflection

In this chapter you will look at three different types of transformation. A transformation is when the position of a shape is changed in some way. The first transformation you will look at is reflection.

Here are some examples of reflection. In each case the dotted line shows the position of the mirror line. The mirror line is a line of symmetry for the diagram.

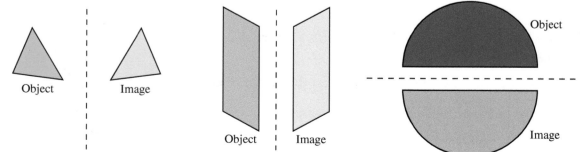

The original shape is called the object and the reflected shape is the image.

In each case the object and the image are the same shape and the same size. They are congruent.

Worked example 1

Copy these shapes on to squared paper. In each case draw the reflection.

a

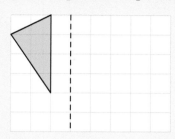

b

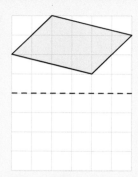

a The image must be the same distance from the mirror line as the object.

This shows how to draw the reflection.

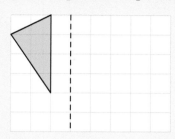

The top point is one square away from the mirror line horizontally.

Mark a point one square away from the line on the other side of the mirror line.

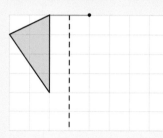

The left hand point is three squares from the line horizontally.

Mark a new point three squares away on the right hand side.

Do the same for the final point.

Join the points up with straight lines as you go.

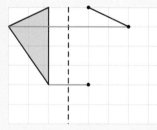

Here is the final image.

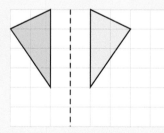

b This is done the same way but the points are vertically above the mirror line.

The reflected points are vertically below the mirror line.

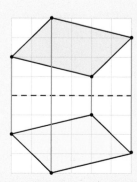

The two mirror lines shown in Worked example 1 lined up with the squares of the paper.

They were either vertical or horizontal. The mirror line may be at an angle.

The next example shows a reflection with a diagonal mirror line.

Worked example 2

Copy this diagram on to squared paper and draw the reflection in the dotted line.

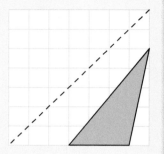

This time check the diagonal distance from each point to the line.

The top point is one square away from the mirror line diagonally.

The red line on the diagram shows this.

The red lines are at right angles to the mirror line.

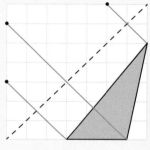

Join the points to get the correct image.

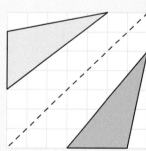

Exercise 13.1

1 Copy these diagrams on to squared paper and draw the reflection of each shape in the mirror line.

a    b    c

d

e

f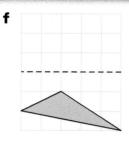

2 Copy these diagrams on to squared paper and draw the reflection of each shape in the mirror line.

a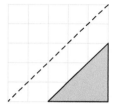

b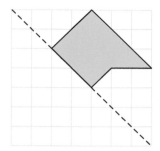

c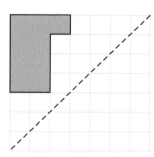

d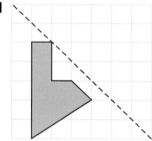

3 Copy these diagrams on to squared paper. In each case draw in the correct mirror line.

a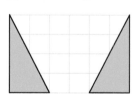

b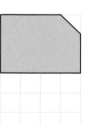

c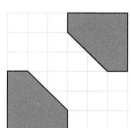

d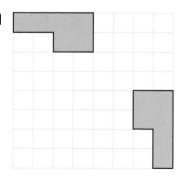

4 Sandra and Alan are both asked to draw the image of an object reflected in a mirror line. Their diagrams are shown below.

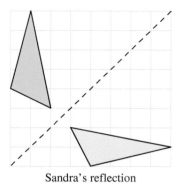

Sandra's reflection

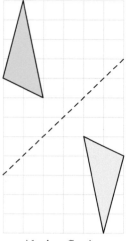

Alan's reflection

Who has drawn the correct image?

5 Here are four statements about reflection. Decide if each statement is true or false.

a The angles of the image are the same as the angles of the object.

b The area of the image is greater than the area of the object.

c The lengths of the sides of the image are the same as the lengths of the sides of the object.

> Look at some diagrams of reflections when you answer this question.

d The image and the object are congruent.

6 Copy this diagram with two diagonal mirror lines. Draw the L-shape and label it *A*.

a Reflect shape *A* in the red mirror line. Label the new shape *B*.

b Reflect shape *A* in the blue mirror line. Label the new shape *C*.

c Reflect shape *B* in the blue mirror line. Label the new shape *D*.

d What happens if you reflect shape *C* in the red mirror line?

> Take care when reflecting in diagonal lines.

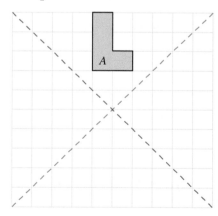

13.2 Rotation

A different type of **transformation** is rotation. A rotation is a turn around a **centre of rotation**. You need the angle of rotation and the direction of the rotation.

Here are three examples of rotation.

1

2

3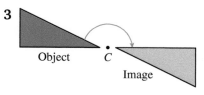
Object   C
Image

This is a clockwise, quarter-turn rotation about the point *A*.

This is an anticlockwise rotation of 135° about the point *B*.

This is a rotation of 180° clockwise about the point *C*.

Note that example 1 uses a fraction of a full turn and the other two examples use degrees.

The original shape is called the **object** and the rotated shape is the **image**.
The object and the image are **congruent**.

Example 3 shows a 180° clockwise rotation. A 180° anticlockwise rotation would have produced the same image.
For 180° rotations there is no need to state the direction.

To draw rotations on plain paper you need to use tracing paper. On squared paper, it is easy to draw rotations of 90° or 180°.

Worked example 1

Rotate this triangle 180° about the point *X*.

X

Place a piece of tracing paper over the shape and the centre of rotation. Trace around the shape in pencil.

X

Hold the centre of rotation with a pin or a sharp pencil point. Gently rotate the tracing paper.

X

The diagram shows an anticlockwise rotation. If you rotate 180° in the opposite direction you should get the same image. Try it in both directions.

Continue rotating until you have rotated the tracing paper 180°.

Draw the finished shape in the correct place.

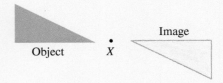

Object   X   Image

Worked example 2

Rotate this shape 90° clockwise about the point *A*.

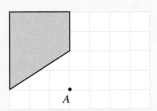

Draw a straight line from one of the corners to the centre of rotation.

Rotate the line through the required turn.

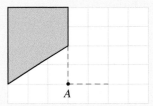

If necessary do the same from the other vertices.

Draw in the completed shape.

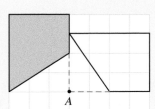

Shade the completed image.

Erase any construction lines.

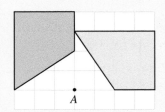

Exercise 13.2

1 Copy these diagrams and draw the image after each rotation.

   a 180° about point *A*

   b 90° anticlockwise about point *B*

   c 90° anticlockwise about point *C*

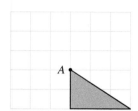 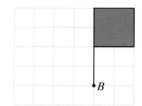 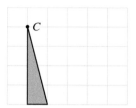

2 Copy these diagrams and draw the image after rotation around centre *X* by the given angle.

   a 90° clockwise

   b 90° clockwise

   c 180°

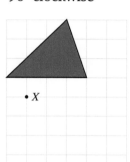 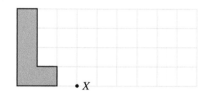 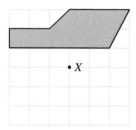

   d 90 ° anticlockwise

   e 180°

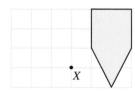 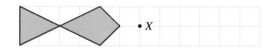

3 Copy this diagram on to squared paper.

   a Rotate the flag 90° clockwise around point *X*.

   b Rotate the new flag 90° clockwise around point *X*.

   c Repeat by rotating again around point *X*.

   d What is the order of rotation symmetry of the final design?

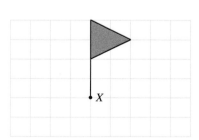

4 This diagram shows some triangles labelled *P*, *Q*, *R*, *S* and *T* and some points labelled *A*, *B*, *C*, *D* and *E*.

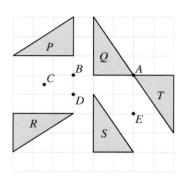

Describe the following rotations:

Use tracing paper to help you find the centres of rotation.

a from *P* to *Q*

b from *P* to *R*

c from *Q* to *T*

d from *T* to *S*

e from *R* to *S*.

5 Copy this diagram on to squared paper.

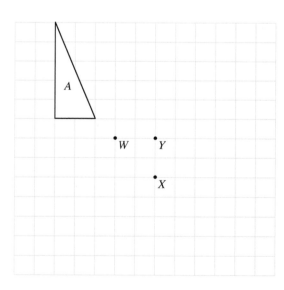

a Rotate triangle *A* 90° clockwise about the point marked *W*. Label the new triangle *B*.

b Rotate triangle *B* 90° clockwise about the point marked *X*. Label the new triangle *C*.

c Rotate triangle *C* 90° clockwise about the point marked *Y*. Label the new triangle *D*.

d Describe the rotation that would take triangle *D* back to triangle *A*.

13.3 Translation

The third **transformation** is translation. A translation is a movement from one place to another without any reflection or rotation. In each case the **object** and the **image** are the same shape and the same size. They are **congruent**.

Here are some examples of translation.

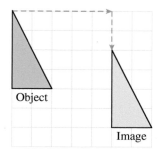

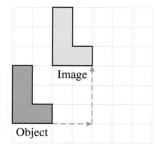

The left hand translation is 5 squares to the right and 2 squares down.

The right hand translation is 2 squares to the right and 3 squares up.

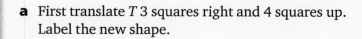

Worked example

The diagram shows a shape, *T*, on squared paper.

a Translate *T* 3 squares to the right and 4 squares up.

   Label the new shape *U*.

b Translate *U* 2 squares to the right and 6 squares down.

   Label the new shape *V*.

c Describe the translation that takes *V* back to *T*.

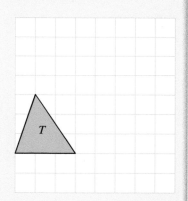

a First translate *T* 3 squares right and 4 squares up.
Label the new shape.

> It helps if you concentrate on one vertex of the triangle at a time. Check that the top vertex has moved 3 squares right and 4 squares up.

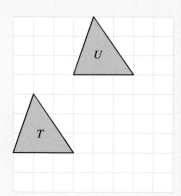

b Now translate U and label the new shape V.

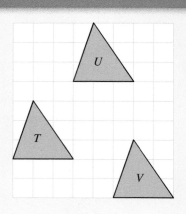

c To work out the translation look at one vertex.

The top of triangle T is 5 squares to the left and 2 squares up from the top of triangle V.

The required translation is 5 squares to the left and 2 squares up.

> You can use any vertex – it does not need to be the top one. Check this by using the bottom left hand vertex.

Exercise 13.3

1 Copy these diagrams on to squared paper and draw the image after the given translation.

a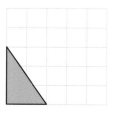

3 squares to the right
2 squares up

b

2 squares to the left
1 square down

c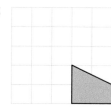

3 squares to the left
3 squares up

d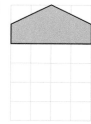

4 squares down

2 A, B, C and D are congruent shapes on a square grid.

Describe the translation in each case:

a from A to B

b from A to C

c from C to D

d from D to A.

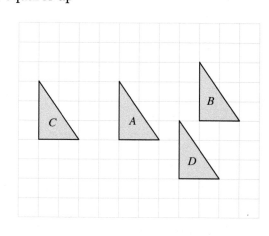

3 Copy this diagram on to squared paper.

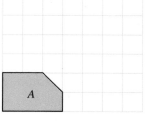

   a Translate shape *A* 1 square to the right and 4 squares up. Label the new shape *B*.

   b Translate shape *B* 2 squares to the right and 1 square down. Label the new shape *C*.

   c Describe the translation that moves shape *A* on to shape C.

4 Copy this diagram on to squared paper.

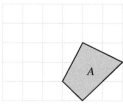

   a Translate shape *A* 3 squares to the left and 2 squares up. Label the new shape *B*.

   b Describe the translation that takes shape *B* on to shape *A*.

5 Copy this diagram on to squared paper.

Use as large a piece of paper as you can.

Draw shape *A* near the left hand side of the paper.

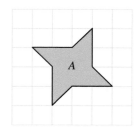

   a Translate the shape 4 squares right and 2 squares down.

      Repeat this translation until you reach the edge of the paper.

   b Translate shape *A* 2 squares right and 4 squares up.

   c Translate your new shape 4 squares right and 2 squares down.

      Repeat this until you reach the edge of the paper.

> If you draw another row of the shapes below this one you can see the white shape more clearly.

   d Carry on with this pattern.

   e What shape are the white gaps that are left between the repeated shapes?

13.4 Using coordinates

A convenient way of describing points is by using **coordinates**.

These are shown on a graph with an *x*-axis and a *y*-axis.

The **axes** are numbered.

The axes cross at the **origin**.

Here is a coordinate grid numbered from −6 to 6 on each axis.

Note the labels on each axis, *x* across and *y* upwards.

There are four points shown on these axes.

The coordinates of a point are given as two numbers in brackets separated by a comma.

$$(x, y)$$

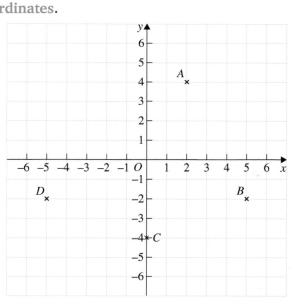

The first number, the x-coordinate, gives the horizontal distance to the left or right of the origin.

If the x-coordinate is negative the point is to the left of the origin.

The second number, the y-coordinate, gives the vertical distance above or below the origin.

If the y-coordinate is negative the point is below the origin.

Note that the coordinates of the origin are $(0, 0)$.

Here are the coordinates of the four points marked on the graph.

The coordinates of A are $(2, 4)$.

The coordinates of B are $(5, -2)$.

The coordinates of C are $(0, -4)$.

The coordinates of D are $(-5, -2)$.

> Note that both numbers are negative because the point D is to the left of the origin and below it.

Worked example 1

Draw a pair of coordinate axes with x-values from -5 to $+5$ and y-values from -5 to $+5$.

Plot the points $(-3, 3)$, $(-3, -2)$ and $(5, -2)$.

These points are at three corners of a rectangle.

Write down the coordinates of the fourth corner.

Here is a grid with the three points plotted.

If you draw a rectangle with these three corners you can see where the other corner must be.

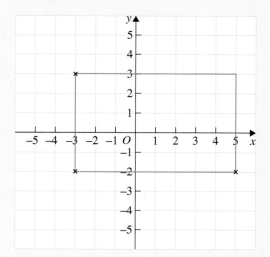

The fourth corner is at $(5, 3)$.

Worked example 2

Draw a pair of coordinate axes with *x*-values from −5 to +5 and *y*-values from −5 to +5.

Draw a triangle with vertices at (1, 2), (1, 5) and (3, 2). Label the triangle *T*.

a Reflect triangle *T* in the *x*-axis. Label the image *V*.

b Rotate triangle *T* 90° anticlockwise about (0, 0). Label the image *W*.

First draw the axes and mark the three vertices of triangle *T*.

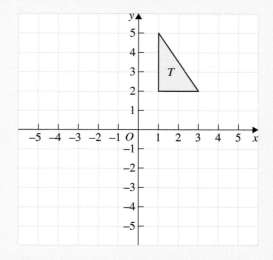

a The *x*-axis runs horizontally. Reflect the triangle in that line as shown in this diagram. Label the image *V*.

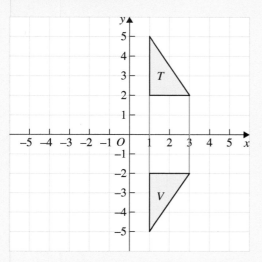

b Rotate triangle *T* anticlockwise 90° about (0, 0).

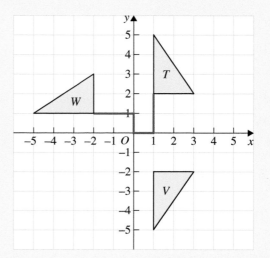

It sometimes helps to think of an L shape from the centre of rotation to the point. An example is shown here in purple.

Exercise 13.4

1 Copy this coordinate grid.

   a Plot the points $(-2, 5)$, $(3, 5)$ and $(3, -4)$.

   These points are at the vertices of a rectangle.

   b Write down the coordinates of the fourth vertex.

2 Draw another copy of the coordinate grid from question **1**.

   a Plot the points $(-4, 0)$, $(2, 0)$ and $(-1, 3)$.

   These points are at the vertices of a square.

   b Write down the coordinates of the fourth vertex.

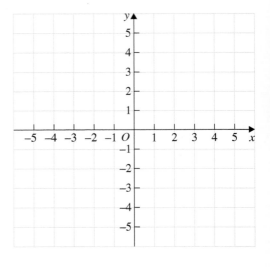

> You need to think diagonally in question **2**.

3 Copy the grid.

Plot the points $(1, 2)$, $(1, 3)$ and $(3, 3)$.
Join the points to make a triangle.

Translate the triangle 2 squares to the right and 3 squares up.

What are the coordinates of the vertices of the new shape?

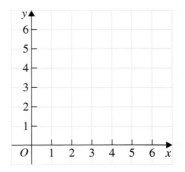

4 Copy the grid.

Plot the points $(1, 1)$, $(1, 4)$, $(2, 4)$ and $(4, 1)$.
Join the points to make a quadrilateral.

Reflect the quadrilateral in the y-axis.

Write down the coordinates of the new shape.

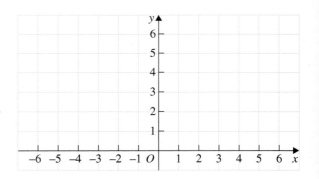

5 The diagram shows a shape labelled A.
Copy the diagram on to squared paper.

The point $(0, 0)$ is marked with a red dot.

   a Rotate shape A 90° anticlockwise about the point $(0, 0)$. Label the new shape B.

   b Reflect shape A in the y-axis. Label the new shape C.

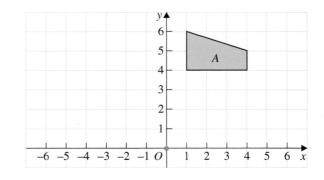

6 Draw axes with x-values from -4 to 4 and y-values from -4 to 4.

Plot the points $(1, 1)$, $(3, 1)$ and $(1, 4)$. Label the triangle A.

a Reflect triangle A in the x-axis. Label the new triangle B.

b Rotate triangle A 180° about the point $(0, 0)$. Label the new triangle C.

c Describe the reflection from triangle B to triangle C.

7 Draw axes with x-values from -5 to 5 and y-values from -5 to 5.

Plot the points $(1, 1)$, $(3, 1)$ and $(1, 5)$. Label the triangle P.

a Reflect triangle P in the y-axis. Label the new triangle Q.

> Take care with the diagrams. Use tracing paper to help with the rotations.

b Describe the transformation that takes Q back to P.

c Rotate triangle P 90° clockwise around the point $(0, 0)$. Label the new triangle R.

d Describe the transformation that takes R back to P.

e Translate triangle P 4 squares to the left and 5 squares down. Label the new triangle S.

f Describe the transformation that takes S back to P.

Review

1 Copy these shapes on to squared paper. Draw the reflections in the given mirror line.

a 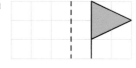 **b** 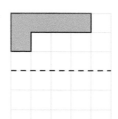 **c**

2 Copy these shapes on to squared paper. Draw the correct mirror line in each one.

a  **b**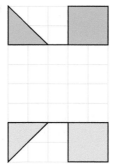

c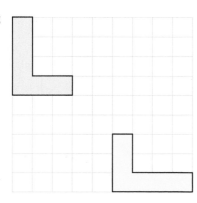

3 Copy these shapes on to squared paper.

In each case draw the image after the given rotation.

a Rotation 90° clockwise around *A*.

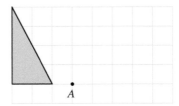

b Rotation 90° anticlockwise around (0, 0).

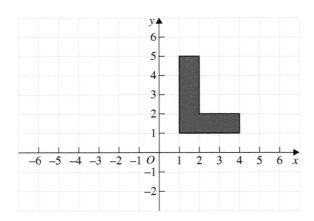

c Rotation 180° around *C*.

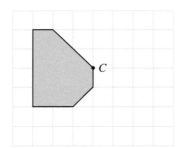

4 This diagram shows shapes *A*, *B*, *C* and *D*.

It also shows three points labelled *X*, *Y* and *Z*.

Describe the following rotations:

a from *A* to *B*

b from *B* to *C*

c from *A* to *D*

The transformation from *C* to *D* is not a rotation.

d Describe the transformation that takes *C* to *D*.

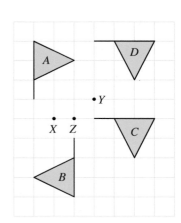

5 Copy these shapes on to squared paper. Describe the translation in each case.

a

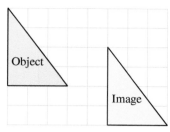

b

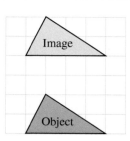

6 a Copy these axes.

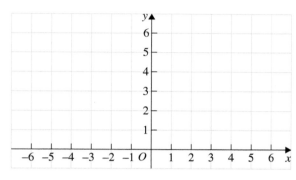

Plot the points $(-5, 5)$, $(-5, 3)$, $(-2, 3)$ and $(-3, 5)$.

Join them up to form a quadrilateral.

Label the shape A.

Translate the quadrilateral 7 squares right and 2 squares down.

Label the new image B.

b Write down the coordinates of the vertices of B.

7 Draw axes with x-values from -4 to 4 and y-values from -4 to 4.

Plot the points $(1, 1)$, $(3, 1)$, $(2, 4)$ and $(1, 4)$. Label the quadrilateral A.

a Rotate shape A clockwise 90° around the point $(0, 0)$. Label the new shape B.

b Reflect shape B in the y-axis. Label the new shape C.

c Write down the coordinates of shape C.

d What type of transformation takes shape A to shape C?

14 Sequences

On the fence

Mathematical patterns appear in all areas of life.

Recognising and understanding how number patterns are formed is an important skill that will help you to solve many practical problems.

If you wanted to build this fence you could use number patterns to work out the number of pieces of wood that are needed.

14.1 Number sequences

A sequence is a list of numbers or diagrams that are connected by a rule.

2, 9, 16, 23, … is a sequence of numbers.

Each number in the sequence 2, 9, 16, 23, … is 7 more than the number before.

The numbers in a sequence are called the terms of the sequence.

The term-to-term rule is 'Add 7.'

Worked example 1

24, 21, 18, 15, …

Write down:　　**a**　the term-to-term rule　　**b**　the next two terms in the sequence.

a　The term-to-term rule is 'Subtract 3.'

b　$15 - 3 = 12$ and $12 - 3 = 9$

　　The next two terms are 12 and 9.

Worked example 2

The first term of a sequence is 4.

The term-to-term rule for the sequence is 'Multiply by 3 and then subtract 2.'

Write down the first four terms of the sequence.

First term   = 4

Second term = 3 × first term − 2 = 3 × 4 − 2 = 10

Third term   = 3 × second term − 2 = 3 × 10 − 2 = 28

Fourth term = 3 × third term − 2 = 3 × 28 − 2 = 82

The first four terms are: 4, 10, 28, 82.

Exercise 14.1

1 For each of these sequences, write down:

    i the term-to-term rule   **ii** the next two terms   **iii** the tenth term.

    a 5, 7, 9, 11, …   **b** 2, 5, 8, 11, …   **c** 7, 12, 17, 22, …

    d 17, 21, 25, 29, …   **e** 34, 47, 60, 73, …   **f** 16, 53, 90, 127, …

    g 30, 28, 26, 24, …   **h** 60, 53, 46, 39, …   **i** 84, 81, 78, 75, …

    j −15, −12, −9, −6, …   **k** 15, 8, 1, −6, …   **l** −9, −13, −17, −21, …

2 Find the missing numbers in these sequences.

    a 8, 14, ☐, 26, 32, ☐, 44, …   **b** ☐, 3, 10, ☐, 24, ☐, …

    c −6, ☐, 2, 6, ☐, ☐, 18, …   **d** ☐, ☐, 34, 29, ☐, ☐, 14, …

    e 20, ☐, ☐, 2, −4, ☐, ☐, …   **f** 32, ☐, 50, ☐, 68, ☐, …

3 In each of these sequences the same number is added each time.

    Find the missing numbers.

    a 4, ☐, ☐, 19, …   **b** 7, ☐, ☐, ☐, 19, …

    c 12, ☐, ☐, ☐, ☐, 52, …   **d** 3, ☐, ☐, ☐, ☐, ☐, 45, …

    e 22, ☐, ☐, ☐, ☐, ☐, ☐, 64, …   **f** 4, ☐, ☐, ☐, 112, …

4 In each of these sequences the same number is subtracted each time.

    Find the missing numbers.

    a 32, ☐, ☐, 26, …   **b** 70, ☐, ☐, ☐, 30, …

    c 9, ☐, ☐, ☐, ☐, ☐, ☐, ☐, −7, …   **d** −6, ☐, ☐, ☐, −30, …

5 For each of these sequences, write down:

   i the next two terms   **ii** the term-to-term rule.

   a 160, 80, 40, 20, …   **b** 2, 6, 18, 54, …   **c** 1, 5, 25, 125, …

   d 729, 243, 81, 27, …   **e** 6, 36, 216, 1296, …   **f** 1, 10, 100, 1000, …

6 Write down the first four terms of each of these sequences.

	First term	Term-to-term rule
a	3	Multiply by 3
b	32	Divide by 2
c	2	Multiply by 2 and then add 1
d	4	Multiply by 3 and then subtract 5
e	1	Add 4 and then multiply by 2
f	2	Subtract 1 and then multiply by 3
g	64	Divide by 2 and then subtract 4
h	10	Add 10 and then divide by 2

7 The second term in a sequence is 27.

   The term-to-term rule is 'Multiply by 2 and then add 5.'

   Find the first term in the sequence.    Solve the equation $2x + 5 = 27$

8 The second term in a sequence is 10.

   The term-to-term rule is 'Divide by 2 and then subtract 3.'

   Find the first term in the sequence.

9 The third term in a sequence is 60.

   The term-to-term rule is 'Add 2 and then multiply by 3.'

   Find the first term in the sequence.

10 Write down the next two terms in each of these algebraic sequences.

   a $x, 2x, 3x, 4x, …$ **b** $x, x + 2, x + 4, x + 6, …$

   c $x, x - 3, x - 6, x - 9, …$ **d** $2x + 1, 2x + 3, 2x + 5, 2x + 7, …$

   e $3x - 4, 6x - 4, 9x - 4, 12x - 4, …$ **f** $20x, 17x, 14x, 11x, …$

   g $5xy, 9xy, 13xy, 17xy, …$ **h** $xy + 2x, 2xy + 4x, 3xy + 6x, 4xy + 8x, …$

   i $x, 2x + 5, 4x + 15, 8x + 35, 16x + 75, …$

11 The sequence 1, 1, 2, 3, 5, 8, 13, 21, … is called a Fibonacci sequence.

   The rule to find the next term in the sequence is 'Add together the last two numbers.'

   (To find the third term you work out $1 + 1 = 2$. To find the fourth term you work out $1 + 2 = 3$)

   Write down the next three terms in the sequence.

12 Write down the next three terms in each of these Fibonacci sequences.

   a 2, 2, 4, 6, 10, …   **b** 5, 6, 11, 17, 28, …

   c 6, −1, 5, 4, 9, …   **d** −9, −9, −18, −27, −45, …

13 47, 42, 22, 18, …

The rule to find the next term in this sequence is 'Multiply together the digits in the last number and then add 14.'

(To find the second term you work out $4 \times 7 + 14 = 28 + 14 = 42$)

   a Find the next four terms in the sequence.

   b What is the 50th number in the sequence?

14.2 Sequences from patterns of shapes

These patterns of shapes form a sequence.

    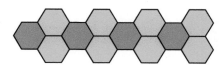

Pattern 1   Pattern 2   Pattern 3   Pattern 4

The number of hexagons in each pattern form the sequence 3, 6, 9, 12, …

The term-to-term rule for the sequence is 'Add 3.'

Put the results in a table to find a rule connecting the number of hexagons with the pattern number.

Pattern number (n)	1	2	3	4
Number of hexagons	3	6	9	12

$\times 3$

Number of hexagons = $3 \times$ pattern number   or   Number of hexagons = $3n$

Worked example

These patterns are made from a number of sticks.

      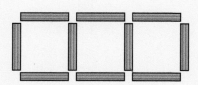

Pattern 1   Pattern 2   Pattern 3

   a Draw the next two patterns in the sequence.

   b Complete the sequence 4, 7, 10, ☐, ☐, …

   c Write down the term-to-term rule.

   d Write down the rule connecting the pattern number and the number of sticks.

   e How many sticks are in pattern 10?

a

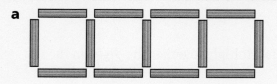
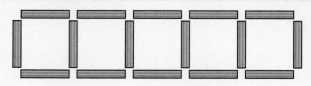

Pattern 4   Pattern 5

b The sequence is: 4, 7, 10, 13, 16, …

c The term-to-term rule is 'Add 3.'

d Put the results in a table to find the rule:

Pattern number (*n*)	1	2	3	4	5
Number of sticks	4	7	10	13	16

× 3 then +1

Number of sticks = 3 × pattern number + 1   or   Number of sticks = $3n + 1$

e Number of sticks in pattern 10 is 3 × 10 + 1 = 31

Exercise 14.2

1

Pattern 1   Pattern 2   Pattern 3

a Draw the next two patterns in the sequence.

b Copy and complete the sequence of dots: 3, 6, 9, ☐, ☐, …

c Write down the term-to-term rule.

d Copy and complete the table.

Pattern number (*n*)	1	2	3	4	5	6
Number of dots	3	6	9			

e Write down the rule connecting the pattern number, *n*, and the number of dots.

f How many dots are in: **i** pattern 10   **ii** pattern 15?

2 These patterns are made from squares.

Pattern 1   Pattern 2   Pattern 3

a Draw the next two patterns in the sequence.

b The number of squares in each pattern forms a sequence.

Copy and complete the sequence: 3, 5, 7, ☐, ☐, …

c Write down the term-to-term rule.

d Copy and complete the table.

Pattern number (*n*)	1	2	3	4	5	6
Number of squares	3	5	7			

e Write down the rule connecting the pattern number, *n*, and the number of squares.

f How many squares are in: **i** pattern 10   **ii** pattern 20?

3 These patterns are made from hexagons.

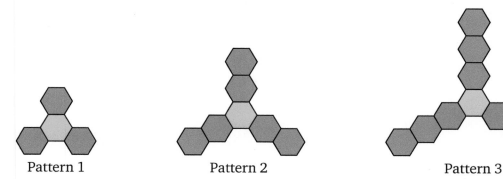

Pattern 1   Pattern 2   Pattern 3

a Copy and complete the table.

Pattern number (n)	1	2	3	4	5	6
Number of hexagons	4	7	10			

b Write down the term-to-term rule.

c Write down the rule connecting the pattern number, n, and the number of hexagons.

d How many hexagons are in: **i** pattern 10   **ii** pattern 20?

4 The number of pieces of wood needed to make a fence form a sequence.

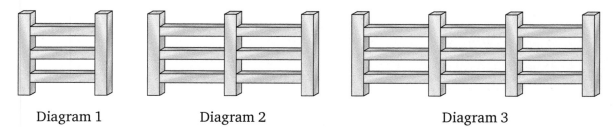

Diagram 1   Diagram 2   Diagram 3

a Copy and complete the table.

Diagram number (n)	1	2	3	4	5	6
Number of pieces of wood	5	9				

b Write down the term-to-term rule.

c Write down the rule connecting the diagram number, n, and the number of pieces of wood.

d How many pieces of wood are in: **i** diagram 15   **ii** diagram 30?

5

Pattern 1   Pattern 2   Pattern 3   Pattern 4

Explain why there are 106 dots in pattern 100. •————

> Use the different coloured dots to help you.

6

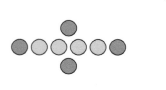

Pattern 1   Pattern 2   Pattern 3

How many dots are in pattern 100? Use the different coloured dots to help you.

7 These steps are made of blocks.

1 step   2 steps   3 steps   4 steps
1 block   3 blocks   6 blocks   10 blocks

How many blocks are needed to make 100 steps?

Review

1 For each of these sequences, write down:

  i the next two terms   **ii** the term-to-term rule.

  a 3, 7, 11, 15, …

  b 2, 9, 16, 23, …

  c −23, −19, −15, −11, …

  d 4, 1, −2, −5, …

  e 3, 6, 12, 24, …

  f 64, 32, 16, 8, …

  g 1, 4, 16, 64, …

  h 10 000, 1000, 100, 10, …

2 In each of these sequences the same number is added each time.

Find the missing numbers.

  a 5, ☐, ☐, ☐, 13, …

  b 13, ☐, ☐, ☐, ☐, 73, …

3 In each of these sequences the same number is subtracted each time.

Find the missing numbers.

a 60, ☐, ☐, 36, …

b 45, ☐, ☐, ☐, 29, …

4 The second term in a sequence is 26.

The term-to-term rule is 'Multiply by 3 and then add 2.'

Find the first term in the sequence.

5 The third term in a sequence is 97.

The term-to-term rule is 'Multiply by 4 and then subtract 3.'

Find the first term in the sequence.

6 These patterns are made from sticks.

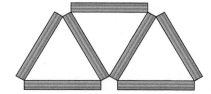

Pattern 1   Pattern 2   Pattern 3

a Copy and complete the table.

Pattern number (n)	1	2	3	4	5	6
Number of sticks	3	5	7			

b Write down the term-to-term rule.

c Write down the rule connecting the pattern number, n, and the number of sticks.

d How many sticks are in: **i** pattern 15   **ii** pattern 30?

7

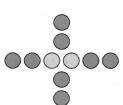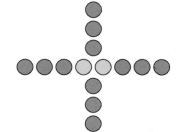

Pattern 1   Pattern 2   Pattern 3

How many dots are in pattern 100?

15 Probability

Is it going to rain?

Everyone thinks about probability at some time.

For example, your decision about whether to wear a raincoat or take an umbrella with you when you go out, depends on whether you think it is likely to rain.

In 1654, two French mathematicians, Blaise Pascal and Pierre Fermat discussed Pascal's ideas on probability. They set up the theory of probability and are considered to be the joint founders of mathematical probability.

15.1 Probability words

Predictions about the future do not always turn out to be correct. It is often a good idea to think about how likely something is to happen.

In the example above, when you go out you may need to think about taking a coat or an umbrella. You may not know whether it is going to rain unless it is raining when you set off, but you may take an umbrella if you think it is likely to rain.

When you think about how likely, or unlikely 'something' is to happen you will use probability words such as impossible, unlikely, even chance, very likely, and certain.

The 'something' is called an event.

You need to be able to match events to these probability words. It is important that you can put the probability words in order.

Impossible	Unlikely	Even chance	Likely	Certain

Worked example

Decide if each of these is certain, likely, unlikely or impossible.

a The next 30 babies born in Wellington will all be girls.

b The day after Wednesday is Monday.

c The sun will rise tomorrow.

d Next month will have 31 days.

a Unlikely.

It is possible that the next 30 babies born in Wellington will all be girls, but it is unlikely.

b Impossible.

The day after Wednesday is always Thursday.

c Certain.

The sun rises every day, even though you may not always see it due to cloud.

d Likely.

Seven months have 31 days, four months have 30 days and one month has 28 or 29 days.

Exercise 15.1

1 Your are asked to carry out a survey of what you can see from your school gates. The survey will take place next Monday and will last for 30 minutes.

Organise the following into five groups, impossible, unlikely, even chance, likely, certain:

baby buggy	van	car	bus	tractor
snowplough	motorbike	bicycle	helicopter	yacht
aeroplane	sledge	train	truck	pedestrian

Copy and complete the table by placing each one in the most suitable column.

Impossible	Unlikely	Even chance	Likely	Certain

2 Sort these statements into four groups.

Use these labels: certain, likely, unlikely, impossible.

a I will have fish for my evening meal tonight.

b An elephant will bring me to school tomorrow.

c Tomorrow will have 24 hours.

d My maths teacher will be the next president.

e I will go shopping next week.

f I will go to bed after 8 o'clock tonight.

g I will be the only person on the bus home from school.

h There will be a good film on television tonight.

i I will not talk to anyone in the next lesson.

j I will be 3 metres tall next year.

k The year 2020 will have 365 days.

l I will get maths homework next week.

m The next Olympic Games will be held in Paris.

3 Kamini has some sweets.

She empties the packs to see how many of each flavour there are.

Her results are shown in the frequency table.

Flavour	Frequency
Lemon	4
Lime	4
Strawberry	3
Pineapple	7
Orange	2

She puts the sweets back in the pack. She shakes the pack. Then she takes a sweet out without looking.

a What flavour is she most likely to take?

b What flavour is she least likely to take?

c Which flavours are equally likely?

d Write the flavours in order of likelihood. Write the most likely flavour first.

4 Arun buys 18 cans of food from the discount store.

They are cheap because they have no labels.

Nine of the cans contain baked beans, five of the cans contain tomatoes, three of the cans contain coconut milk and one of the cans contains peaches.

Arun takes a can.

Decide whether each of the following statements is true (T) or false (F).

a The can is most likely to contain baked beans.

b The can is least likely to contain coconut milk.

c The can is more likely to contain tomatoes than peaches.

d The can is equally likely to contain baked beans as something else.

e The can is more likely to contain coconut milk or peaches than tomatoes.

5 Jamie has eight socks in four colours in a drawer (grey, black, blue and red).

He takes one sock from the drawer without looking.

Use the following statements to decide how many socks of each colour he has.

- The sock is more likely to be grey than black.
- The sock is as likely to be blue as black.
- The sock is least likely to be red.

> A colour is less likely to be chosen if there is a smaller number of that colour in the drawer.

15.2 Probability scale and calculating probabilities

Probability scale

Probability is measured on a scale from 0 to 1.

A probability of 0 means that the event is impossible – there is no chance of it happening. For example, the probability you will visit the Moon tomorrow is 0.

A probability of 1 means the event is certain to happen. For example, if you drop a coin the probability it will fall downwards is 1.

A probability of $\frac{1}{2}$ means there is an even chance of the event happening. For example, if you drop a coin the probability it will land showing a 'tail' is $\frac{1}{2}$

You can show probabilities on a probability scale.

> ### Worked example 1
>
> Rosa has four pairs of socks in her drawer. Two pairs are black, one pair is grey and one pair is white.
>
> She takes a pair of socks from the drawer at random. •——('At random' means that each pair is equally likely to be chosen.
>
> Place an arrow on a probability scale to show the probability of each of these events.
>
> **A** The pair of socks chosen is black.
>
> **B** The pair of socks chosen is blue.
>
> **C** The pair of socks chosen is one of black, grey or white.
>
>
>
> **A** Half the pairs of socks are black. There is an even chance of this happening, so the probability is $\frac{1}{2}$
>
> **B** There are no blue socks in the drawer. There is no chance of the socks being blue, so the probability is 0.
>
> **C** All the socks are either black, grey or white. The socks are certain to be one of these colours, so the probability is 1.

Equally likely outcomes

When two or more results or outcomes have the same probability they are called equally likely outcomes.

For example a fair coin is as likely to land on 'heads' as it is 'tails'.

A fair dice is just as likely to land on a 1 as a 2, or a 3 as a 4, or a 5 as a 6.

These are examples of equally likely outcomes.

When equally likely outcomes are used to find a probability it is called a theoretical probability.

Calculating probabilities

This formula is used to find the probability of an outcome occurring.

$$\text{Probability of a successful outcome} = \frac{\text{number of successful outcomes}}{\text{total number of outcomes}}$$

You may need to list all the successful outcomes and all the possible outcomes in order to find out how many of each there are.

Probabilities can be written as fractions, decimals or percentages. Fractions or decimals are usually best.

Worked example 2

Tracey has to choose a student from her class to work with. To make it fair she numbers them from 1 to 20. Then she chooses one of the numbers at random.

What is the probability the number she chooses is a multiple of 3?

There are 20 possible outcomes: 1, 2, 3, 4, 5, 6, 7, 8, 9, 10, 11, 12, 13, 14, 15, 16, 17, 18, 19, 20

There are 6 successful outcomes: 3, 6, 9, 12, 15, 18

$$\text{Probability of a successful outcome} = \frac{\text{number of successful outcomes}}{\text{total number of outcomes}}$$

$$\text{Probability of a multiple of 3} = \frac{6}{20} = \frac{3}{10}$$

'Probability of a multiple of 3' is often written as P(multiple of 3).

Exercise 15.2

1 There are four biscuits on a plate. Two are lemon biscuits, one is a ginger biscuit and one is a chocolate biscuit. Bella takes one of the biscuits at random.

Place an arrow on a probability scale to show the probability of each of these events.

A The biscuit is lemon.
B The biscuit is ginger.

C The biscuit is a cheese biscuit.
D The biscuit is lemon or ginger or chocolate.

2 A fair dice is thrown once.

What is the probability of getting:

a a six
b an odd number

c a number less than 5
d a number greater than 6

e a number greater than 0?

3 The diagram shows a probability scale. The arrows point to different probabilities.

An ordinary fair dice is rolled.

a Which arrow points to the probability of rolling a 3?

b Which arrow points to the probability of rolling a number greater than 1?

c Draw an arrow on the scale to show the probability of rolling a multiple of 3.

4 There are 12 discs numbered from 1 to 12. A student places them into a bag and then takes one out of the bag without looking.

What is the probability the disc:

a has the number 5 on it

b has an even number on it

c has a number less than 10 on it

d has a number greater than 9 on it

e has a number that is a multiple of 5 on it?

5 The 11 letters of the word PROBABILITY are written on separate cards. The cards are mixed up.

Matt chooses one of the cards without looking.

What is the probability that the letter on it is:

a the letter Y

b the letter B

c a vowel?

The vowels are A, E, I, O and U.

6 Here is a list of numbers.

3,   5,   3,   4,   2,   7,   6,   8,   3

One of the numbers is chosen at random.

a What is the probability the number chosen is 3?

b Put these events in order of likelihood. Start with the least likely.

A The number is 3.

B The number is even.

C The number is greater than 6.

7 Kamini has some sweets. She empties the packs to see what flavours are there.

Her results are shown in the frequency table.

Flavour	Frequency
Lemon	4
Lime	4
Strawberry	3
Pineapple	7
Orange	2

She puts the sweets back in the pack. She shakes the pack. She takes one out without looking.

What is the probability the sweet is:

a strawberry flavour

b pineapple flavour

c orange flavour

d lemon or lime flavour

e not orange in flavour?

8 Gita likes orange sweets, but not blue ones.

She can choose a sweet from Bag A or Bag B.

Which bag should she choose?

Give a reason for your answer.

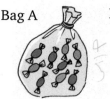

Bag A   Bag B

9 There are blue, black and white socks in a drawer.

The probability of getting a blue sock out of the drawer is 0.25

The probability of getting a black sock is 0.5

There are 10 white socks.

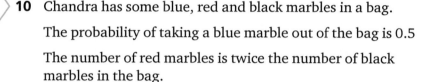

> Start by working out the probability of getting a white sock.

   a How many blue socks are there?

   b How many black socks are there?

10 Chandra has some blue, red and black marbles in a bag.

The probability of taking a blue marble out of the bag is 0.5

> Use fractions to work out the probability of taking a red marble out of the bag.

The number of red marbles is twice the number of black marbles in the bag.

There are 5 more blue marbles than red marbles in the bag.

How many marbles are in the bag altogether?

11 There are six balls in a bag.

The probability of taking a red ball from the bag is 0.5

A red ball is removed from the bag.

> First find the number of red balls in the bag using the probability given.

What is the probability of taking a red ball from the bag now?

12 Boxes of cereal contain cards to be collected. Each box contains one card. Each card is equally likely to be in any one box.

Four different cards are needed to make a set.

It is not possible to know what card is in the box until the box is opened.

Lola needs card A to make a set. Toni needs cards B and C to make a set.

They have a new box of cereal.

   a What is the probability the box contains the card that Lola needs?

   b What is the probability the box contains a card that Toni needs?

They open the box. The card inside is not card A.

   c What is the probability now that it is the card that Lola needs?

   d What is the probability now that it is a card that Toni needs?

13 Paz has seven red marbles and five blue marbles in a bag. He takes a marble at random from the bag, notes the colour and then replaces the marble in the bag.

   a What is the probability it is blue?

  b What is the probability it is not blue?

  c Paz adds six black marbles to the marbles in the bag.

  He says 'the probability of taking a black marble from the bag is $\frac{6}{12}$'

  Explain why he is wrong.

  d What is the probability of taking a black marble from the bag?

> Start by working out the total number of marbles in the bag.

15.3 Estimating probabilities

Sometimes it is not possible to know whether all possible outcomes are equally likely. For example, when you drop a drawing pin there are two possible outcomes – it may land point up or point down. These two outcomes may, or may not, be equally likely.

You could do an **experiment** to investigate.

You could drop a drawing pin lots of times and record the results.

Each time you drop the drawing pin is called a **trial**.

You can use the result of the experiment to find the **relative frequency** of the drawing pin landing point up. This is the number of times the drawing pin lands point up as a fraction of the total number of times you drop the drawing pin.

$$\text{relative frequency} = \frac{\text{number of successful outcomes}}{\text{total number of trials}}$$

In this experiment, landing point up is the successful outcome, so:

$$\text{relative frequency} = \frac{\text{number of times the drawing pin lands point up}}{\text{total number of times you drop the drawing pin}}$$

The relative frequency gives you an estimate of the probability that the drawing pin will land point up.

This is called the **experimental probability** because it comes from carrying out an experiment.

Worked example 1

A company makes light bulbs. They test a sample of 100 bulbs to find out how long they last. The frequency table shows the results.

Lifetime of light bulb	Frequency
Less than 1000 hours	5
From 1000 hours to less than 2000 hours	36
From 2000 hours to less than 3000 hours	49
3000 hours or more	10

What is the probability a light bulb chosen at random will last

  a less than 1000 hours

  b 3000 hours or more

  c 2000 hours or more?

a Relative frequency of less than 1000 $= \frac{5}{100} = \frac{1}{20}$

So, P(less than 1000) $= \frac{1}{20}$

b Relative frequency of 3000 or more $= \frac{10}{100} = \frac{1}{10}$

So, P(3000 hours or more) $= \frac{1}{10}$

c Relative frequency of 2000 or more $= \frac{49 + 10}{100} = \frac{59}{100}$

So, P(2000 hours or more) $= \frac{59}{100}$

Worked example 2

Every day Emilio chooses a piece of fruit from his fruit bowl.

At the start of every week the fruit bowl contains 8 apples, 6 pears, 6 bananas, 8 oranges and 2 kiwi fruits.

One week Emilio chooses 6 apples and 1 kiwi fruit.

Do you think Emilio chooses the fruit at random from the fruit bowl? Give a reason for your answer.

No, the fruit was not chosen at random.

Apples and oranges are both equally likely to be chosen, but Emilio does not choose any oranges.

There are 12 pears and bananas combined, so it is likely that he would choose a pear or banana at least once.

After 3 days of choosing an apple there are only 5 left, so it is more likely that a pear, banana or an orange would be selected than another apple.

Exercise 15.3

1 Adisa has a large bag of coloured counters.

She takes one out and records its colour. She then places the counter back in the bag.

She repeats this 50 times. Her results are shown in the frequency table.

Colour	Frequency
Red	12
Blue	15
Green	19
Yellow	4

Estimate the probability that a counter taken from the bag at random will be:

a red **b** green **c** blue.

2 A bottle contains red beads and blue beads.

Chima chooses a bead from the bottle, records its colour then replaces the bead.

His results are recorded in the frequency table.

Colour	Frequency
Red	33
Blue	67

a How many times did Chima repeat the trials?

b A bead is taken from the bottle at random. What is the probability that it is:

i red

ii blue?

c What does this tell you about the number of red and blue beads in the bottle?

3 A six-sided dice has the numbers 2, 3, 3, 4, 4, 4 on its faces.

Nkiru rolls the dice 50 times and records these results.

Result	Frequency
2	6
3	16
4	28

a What is the theoretical probability of getting a 2 when the dice is rolled?

b What is the experimental probability of getting a 2 on this dice?

c What is the theoretical probability of getting a 3 when the dice is rolled?

d What is the experimental probability of getting a 3 on this dice?

4 Lily, Emma and Sam each rolled a fair dice 240 times.

Here are the results they gave their teacher.

Number	Lily	Emma	Sam
1	42	25	40
2	37	34	40
3	41	28	40
4	36	96	40
5	43	34	40
6	41	23	40

One of them has recorded their results accurately but the other two have not.

Whose results are recorded accurately? Give reasons for your answer.

Think about the results you would expect to get if you rolled a fair dice 240 times.

Review

1 Match each of these words to a statement below.

Impossible   Unlikely   Likely   Certain

a I will eat a packet of crisps today.

b Next year will have 55 Mondays.

c I will leave the classroom through the door.

d The sun will rise in the east tomorrow.

e I will have a sandwich for my lunch today.

f I will meet the president of America on my way home today.

2 The letters of the word RABBIT are written on 6 cards, one letter on each card.

The cards are mixed up and then placed on a table with the letters facing down.

One card is turned over.

Place an arrow on a probability scale to show the probability of each of these events.

A The letter on the card is R.

B The letter on the cards is B.

C The letter on the card is S.

3 20 discs numbered from 1 to 20 are placed into a bag.

One is taken out of the bag without looking.

What is the probability the disc:

a has just the number 5 on it

b has an odd number on it

c has a number less than 10 on it

d has a number greater than 14 on it

e has a number which is a multiple of 5 on it

f has a number which is prime on it

g has a number with the figure 1 in it?

4 A paper boy delivers these newspapers, one to each house.

Times	15
Post	3
Express	8
Mail	4

One of the houses he delivers to is chosen at random.

What is the probability that this house has:

a the *Times*

b the *Post*

c either the *Express* or the *Mail*?

5 Kwesi rolls a dice 100 times.

Here are his results.

Score	Frequency
1	22
2	22
3	9
4	18
5	8
6	21

Do these results suggest the dice is fair or not fair?

Give a reason for your answer.

16 Functions and graphs

The use of graphs

A graph is a useful way of showing the relationship between two variables.

Graphs are often used to show how a variable changes with time.

It is much easier for you to see what is happening to a variable by looking at a graph instead of looking at a table of data.

In this chapter you will learn about functions and straight-line graphs (linear graphs).

16.1 Functions

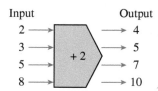

This diagram shows a function machine.

A function is a rule connecting two sets of numbers.

This function machine adds 2 to the numbers that are put into the machine.

The numbers that are put into the function machine are called the input.

The numbers that come out of the number machine are called the output.

Some function machines perform more than one operation.

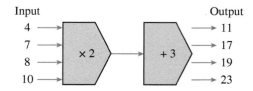

This function machine multiplies by 2 and then adds 3.

The function can also be shown as a mapping diagram:

You say: 4 maps to 11, 7 maps to 17, 8 maps to 19 and 10 maps to 23.

The rule connecting the input and output can be written algebraically as:

$x \rightarrow 2x + 3$ • You say: x maps to $2x + 3$

Worked example 1

Find the missing numbers in this function machine.

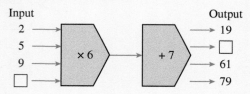

When the input is 5:

Output $= 5 \times 6 + 7 = 37$

When the output is 79:

Solve $6x + 7 = 79$ to find the input.

$6x + 7 = 79$ subtract 7 from both sides

$6x = 72$ divide both sides by 6

$x = 12$

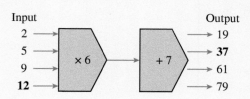

Worked example 2

Find the missing numbers in this mapping diagram.

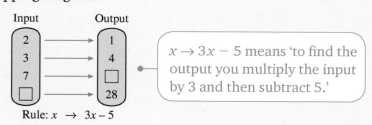

Rule: $x \rightarrow 3x - 5$

$x \rightarrow 3x - 5$ means 'to find the output you multiply the input by 3 and then subtract 5.'

When the input is 7:

Output = $3 \times 7 - 5 = 16$

When the output is 28:

Solve $3x - 5 = 28$ to find the input.

$3x - 5 = 28$ add 5 to both sides

$3x = 33$ divide both sides by 3

$x = 11$

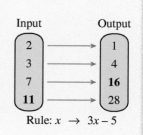

Exercise 16.1

1 Copy and complete these function machines.

a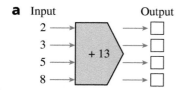

b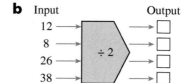

c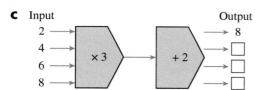

d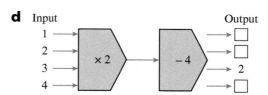

e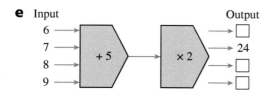

f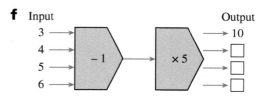

2 Copy and complete these function machines.

a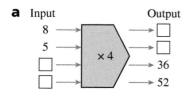

b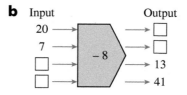

c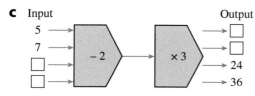

d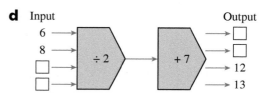

e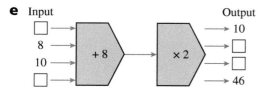

f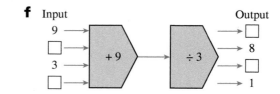

3 a Find the rule to complete this function machine.

b Write down the term-to-term rule for the sequence of outputs 30, 36, 42, 48.

c What is the connection between the rule for the function and the term-to-term rule for the outputs?

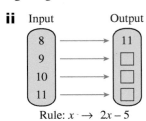

4 a Copy and complete these mapping diagrams.

i
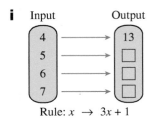
Rule: $x \rightarrow 3x + 1$

ii
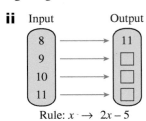
Rule: $x \rightarrow 2x - 5$

iii
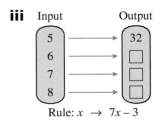
Rule: $x \rightarrow 7x - 3$

b Look at your mapping diagrams for part **a**.

What is the connection between the rule for the mapping diagram and the term-to-term rule for the outputs?

5 Find the rule for each of these mapping diagrams.

Write your answer in the form $x \rightarrow \ldots$

a
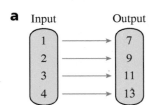

b
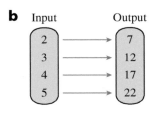

c
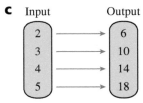

6 Copy and complete these mapping diagrams.

a
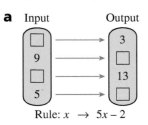
Rule: $x \rightarrow 5x - 2$

b
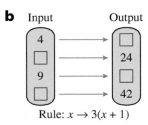
Rule: $x \rightarrow 3(x + 1)$

c
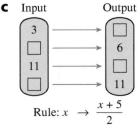
Rule: $x \rightarrow \dfrac{x + 5}{2}$

7 Kiran, Sarah and Jacques are asked to find the rule to complete this function machine.

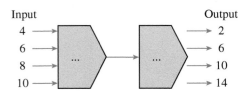

Kiran says: I think the rule is 'Multiply by 2 and then subtract 2.'

Sarah says: I think the rule is 'Multiply by 2 and then subtract 6.'

Jacques says: I think the rule is 'Subtract 3 and then multiply by 2.'

Who is correct? Explain your answer.

16.2 Horizontal and vertical lines

The points (2, 5), (2, −3), (2, 2) and (2, −4) are plotted on the grid.

The points are in a straight line.

The x-coordinate is 2 at every point on this line.

The equation of the line is $x = 2$.

You can describe the line $x = 2$ as a vertical line crossing the x-axis at 2.

Similarly, the line $x = -4$ is a vertical line crossing the x-axis at −4.

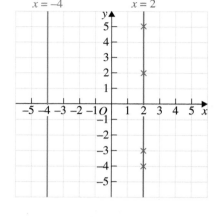

The points (3, 4), (0, 4), (−4, 4) and (2, 4) are plotted on the grid.

The points are in a straight line.

The y-coordinate is 4 at every point on this line.

The equation of the line is $y = 4$.

You can describe the line $y = 4$ as a horizontal line crossing the y-axis at 4.

Similarly, the line $y = -3$ is a horizontal line crossing the y-axis at −3.

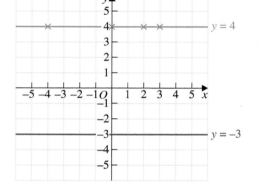

The x-axis is the line $y = 0$

The y-axis is the line $x = 0$

Worked example

Find the equation of the line through the points (2, 5) and (−3, 5).

The points (2, 5) and (−3, 5) both have a y-coordinate of 5.

The equation of the line is $y = 5$.

Exercise 16.2

1 Plot the points (5, 2), (−3, 2), (1, 2), (−4, 2) and (0, 2) on a graph.

Write down the equation of the line through these points.

2 Plot the points (−3, 2), (−3, 1), (−3, −2), (−3, 0) and (−3, −5) on a graph.

Write down the equation of the line through these points.

3 Write down the equations of each of these lines

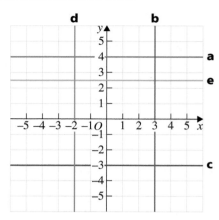

4 Write down the equations of the lines of symmetry of each of these shapes.

a

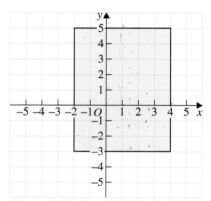

b

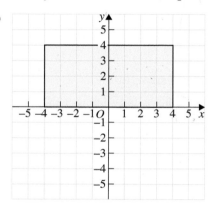

c

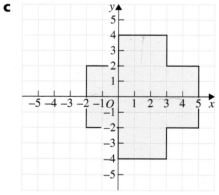

d

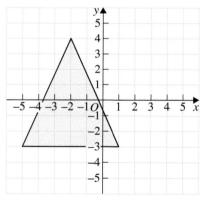

e

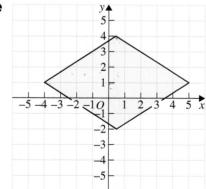

f

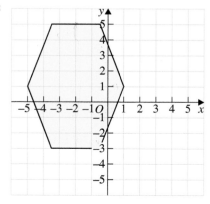

5 Find the equation of the line through each pair of points.

   a (5, 7) and (1, 7)   **b** (4, 8) and (4, 2)   **c** (3, −5) and (10, −5)

   d (−1, 6) and (−1, 7)   **e** (0, 15) and (0, −12)   **f** (−9, −9) and (9, −9)

6 On a coordinate grid draw a shape that has $x = 2$ and $y = 3$ as lines of symmetry.

7 The vertices of a kite are at the points A (2, 4), B (−1, 5), C (−4, 4) and D (−1, −1).

   a Draw the kite on a coordinate grid.

   b The kite has two diagonals.

     Write down the equations of the two diagonals.

8   (3, 2)   (2, 2)   (−3, 2)   (−2, −3)   (−2, 2)

   Find the odd one out. Explain your answer.

9   (5, 3)   (3, 5)   (−5, −5)   (−3, 3)

     (3, −5)   (−5, −3)   (−3, 5)   (3, −3)

   Three of these points are in a straight line.

   Find the equation of the line through the three points.

16.3 Other straight lines

The graph of $y = 3x - 1$ is also a straight line.

To draw the graph you must choose three values for x and substitute them into the equation to find the y-values.

When $x = 0, y = 3 \times 0 - 1 = -1$

When $x = 1, y = 3 \times 1 - 1 = 2$

When $x = 2, y = 3 \times 2 - 1 = 5$

Record your results in a table and then draw the graph.

x	0	1	2
y	−1	2	5

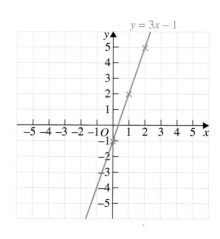

Worked example

Draw the graph of $y = \frac{1}{2}x + 2$

When $x = 0$, $y = \frac{1}{2} \times 0 + 2 = 2$

When $x = 2$, $y = \frac{1}{2} \times 2 + 2 = 3$

When $x = 4$, $y = \frac{1}{2} \times 4 + 2 = 4$

x	0	2	4
y	2	3	4

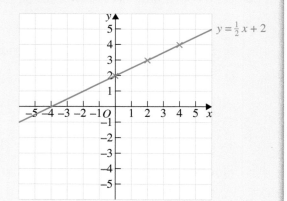

Plot your points clearly with a cross.

Use a ruler and pencil to draw the straight line.

Extend the line to the edge of the grid.

Exercise 16.3

1 a Copy and complete this table of values for $y = x + 2$

$y = x + 2$ means 'to find the y number you add 2 to the x number.'

x	−5	1	3
y			5

b Copy the coordinate grid and draw the graph of $y = x + 2$

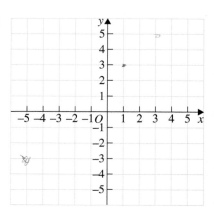

2 You only need two points to draw a straight-line graph.

Why is a third point usually used?

3 For each of the following, copy and complete the table of values and then draw the graph.

Use coordinate axes with values from −5 to 5.

a $y = x$

x	−3	2	4
y			

b $y = x + 4$

x	−2	0	1
y			

c $y = x - 2$

x	1	3	5
y			

4 **a** Copy and complete this table of values for $y = 2x + 1$ •——

> $y = 2x + 1$ means 'to find the y number you multiply the x number by 2 and then add 1.'

x	0	1	2
y			

b Copy the coordinate grid and draw the graph of $y = 2x + 1$

Use coordinate axes with values from -5 to 5.

5 For each of the following, copy and complete the table of values and then draw the graph.

Use coordinate axes with values from -5 to 5.

a $y = 2x - 1$

x	1	2	3
y			

b $y = 3x - 2$

x	0	1	2
y			

c $y = \frac{1}{2}x + 1$

x	0	2	4
y			

6 **a** Copy and complete this table of values for $y = 4 - x$ •——

> $y = 4 - x$ means 'to find the y number you subtract the x number from 4.'

x	0	1	2
y			

b Copy the coordinate grid used for question **1b** and draw the graph of $y = 4 - x$.

7 For each of the following, copy and complete the table of values and then draw the graph.

Use coordinate axes with values from -5 to 5.

a $y = 5 - x$

x	0	1	2
y			

b $y = 3 - x$

x	0	2	4
y			

c $y = 1 - x$

x	0	3	5
y			

8

(3, 11)   (4, 12)   (1, 7)   (2, 10)   (5, 15)

Which of these points lie on the line $y = 2x + 5$?

9

(10, 2)   (4, −2)   (6, 0)   (12, 4)   (8, 1)

Which of these points lie on the line $y = \frac{1}{2}x - 3$?

10 **a** Copy and complete the table of values for the graph shown.

x	0	1	2	3
y				

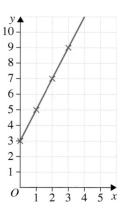

b Find a rule connecting the x- and y-coordinates.

Write the rule as an equation.

11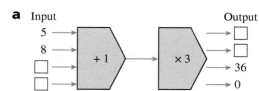

$(3, 10)$ $(7, 16)$ $(0, 4)$ $(10, 24)$ $(2, 8)$

Four of these points are on a straight line.

a Write down the equation of the line.

b Which point is not on the line?

Review

1 Copy and complete these function machines.

a Input → $+1$ → $\times 3$ → Output

Input: 5, 8, □, □
Output: □, □, 36, 0

b Input → -3 → $\div 2$ → Output

Input: 11, □, 15, □
Output: □, 8, □, 1

2 Find the rule for each of these mapping diagrams.

Write your answer in the form $x \to \ldots$

a
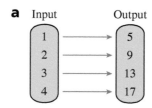

Input	Output
1	5
2	9
3	13
4	17

b
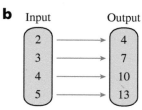

Input	Output
2	4
3	7
4	10
5	13

c
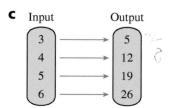

Input	Output
3	5
4	12
5	19
6	26

3 Write down the equations of each of these lines.

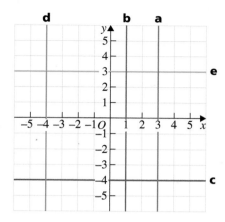

4 Write down the equations of the lines of symmetry of this shape.

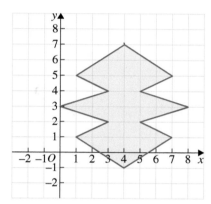

5 For each of the following, copy and complete the table of values and then draw the graph.

Use coordinate axes with values from −5 to 5.

a $y = x + 1$

x	−3	2	4
y			

b $y = 2x - 2$

x	0	1	3
y			

c $y = \frac{1}{2}x - 1$

x	0	2	4
y			

6

(3, 12)　　(10, 26)　　(6, 32)　　(7, 20)　　(1, 9)

Which of these points lie on the line $y = 2x + 6$?

17 Fractions, decimals and percentages

'By the hundred'

'Per cent' means 'out of every hundred' and comes from the Latin *per centum* meaning 'by the hundred'. 'Cent' is a word that is commonly used in systems of measure where there are one hundred smaller units to one larger unit. Examples are cents in a dollar and centimetres in a metre.

The term 'per cent' is usually shortened to %.

17.1 Describing part of an amount

Using fractions to describe part of an amount

Fractions can be used to describe part of an amount.

Remember that the number on the bottom of the fraction is called the denominator. It tells you how many equal parts the whole has been divided into. The number on the top of the fraction is called the numerator. It tells you how many of the equal parts you have, or are interested in.

Worked example 1

What fraction of this rectangle is shaded?

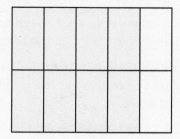

The rectangle is divided into ten equal parts.

Eight of the parts are shaded.

So $\frac{8}{10}$ of the shape is shaded.

In Chapter 4 you learned that $\frac{8}{10}$ can be simplified to $\frac{4}{5}$

$\frac{8}{10}$ and $\frac{4}{5}$ are called **equivalent fractions**.

When you write $\frac{8}{10}$ as $\frac{4}{5}$, this is called writing the fraction in its simplest form, or its lowest terms.

Worked example 2

Max has 12 crayons in his pencil case.

Three crayons are red, four are blue, one is black, one is yellow, two are green and one is purple.

What fraction of Max's crayons are:

a black

b green

c blue

d red?

Give your answers in their simplest forms.

a 1 out of 12 crayons are black. This is written as $\frac{1}{12}$

b 2 out of 12 crayons are green. This is written as $\frac{2}{12} = \frac{1}{6}$

c 4 out of 12 crayons are blue. This is written as $\frac{4}{12} = \frac{1}{3}$

d 3 out of 12 crayons are red. This is written as $\frac{3}{12} = \frac{1}{4}$

Using percentages to describe part of an amount

Percentages can be used to describe part of an amount in the same way as fractions.

You can think of a percentage as being a fraction out of 100.

When a shape is divided into 100 equal parts you only need to count the number of shaded parts to find the percentage that is shaded. Each part is 1% of the whole.

When a shape is not divided into 100 parts, count the number of shaded parts to find the fraction of the shape that is shaded. Then find the equivalent fraction that has denominator 100. This gives the percentage of the shape that is shaded.

> ### Worked example 3
>
> What percentage of these shapes is shaded?
>
> a 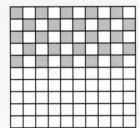   b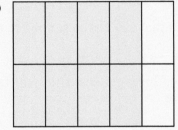
>
> **a** The large square is divided into 100 equal small squares.
>
> The number shaded is 25.
>
> Each shaded square is 1%.
>
> So, 25% of the large square is shaded.
>
> **b** $\frac{8}{10}$ of the shape is shaded.
>
> > Remember that in order to find equivalent fractions you multiply (or divide) the numerator and denominator by the same whole number. In this example you need to multiply the denominator by 10 to get 100. You also need to multiply the numerator by 10 to find the equivalent fraction.
>
> $$\frac{8}{10} = \frac{80}{100}$$
>
> So, 80% of the shape is shaded.

Worked example 4

Paula has a tin of kidney beans.

The label on the tin says:

What percentage of the beans are:

100 g contains

Protein 7 g
Fat 1 g
Fibre 6.2 g

a protein

b fat

c fibre

d not protein, fat or fibre?

a 7 g out of 100 g are protein. 7 out of 100 parts are protein. So 7% of the beans are protein.

b 1 g out of 100 g is fat. So 1% of the beans are fat.

c 6.2 g out of 100 g are fibre. So 6.2% of the beans are fibre.

d 7 g + 1 g + 6.2 g = 14.2 g

So 14.2% is protein, fat or fibre.

100 g − 14.2 g = 85.8 g

So 85.8% of the beans are not protein, fat or fibre.

Exercise 17.1

You must not use a calculator in this exercise.

1 Write down the fraction shaded for each of the following shapes.

Write each fraction in its simplest form.

a

b

c

d

e

f

2 Louisa has 15 books on her bookshelf.

7 are novels, 3 are comedies and 5 are thrillers.

Give your answer for each of the following in its simplest form.

a What fraction are novels?

b What fraction are comedies?

c What fraction are thrillers?

d What fraction are not novels?

e What fraction are not thrillers?

f What fraction are not comedies?

3 Write down the percentage that is shaded for each of the following shapes.

a 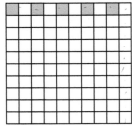   **b**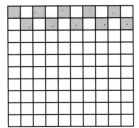

c 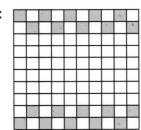   **d**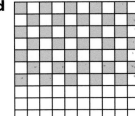

e 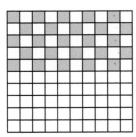   **f**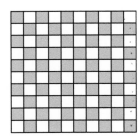

g 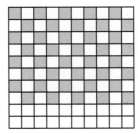   **h**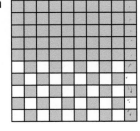

4 For each of the squares in question **3** write down the fraction that is shaded.

Give each answer in its simplest form.

5 Use your answers to question **3** and **4** to complete the following table showing equivalent fractions, decimals and percentages.

The first line has been completed for you.

> Remember to change a fraction to a decimal you divide the numerator by the denominator.

Fraction	Decimal	Percentage
$\frac{1}{20}$	0.05	5%
$\frac{1}{10}$		
	0.2	20%
	0.25	
		35%
$\frac{2}{5}$		
	0.5	
$\frac{3}{4}$		

6 Abdou loves to eat cheese.

Here is the fat content of some of his favourite cheeses:

9 g fat in every 25 g of cheddar

6 g fat in every 25 g of feta

3 g fat in every 25 g of ricotta

1.4 g fat in every 5 g of parmesan

1.1 g fat in every 50 g of cottage cheese

Find the percentage of fat in each of these cheeses.

7 In Sukai's maths class 52% of the students are female.

What percentage of the students are male?

8 Kebba spends 25% of his weekly pocket money during the week. He spends 45% of his pocket money at the weekend. He saves the rest.

What percentage of his pocket money does Kebba save?

9 At a football match 87% of the seats are used.

What percentage of the seats are not used?

10 Sarah takes tests in three subjects. Her test marks are:

78% in maths

15 out of 20 in geography

20 out of 25 in history.

In which subject did Sarah do best?

What was her percentage in this subject?

11 Chloe wants to find the percentage of fat in some foods.

She records the mass of the food and the fat content in a table.

Copy the table and complete the final column.

Food	Mass	Fat content	Percentage of fat
Item 1	25 g	4 g	
Item 2	40 g	18 g	
Item 3	20 g	3 g	
Item 4	80 g	32 g	
Item 5	60 g	27 g	
Item 6	90 g	36 g	
Item 7	15 g	6 g	
Item 8	12 g	9 g	

12 Wei has a pack of 24 sweets.

He gives 25% of the sweets to his brother. Then he eats $\frac{1}{3}$ of the rest himself. He gives the remaining sweets to his sister.

What percentage of the pack does he give to his sister?

> Start by finding 25% of 24. Then work out $\frac{1}{3}$ of your answer.

17.2 Finding and using percentages of an amount

Finding a percentage of an amount

To find a percentage of an amount you first write the percentage as a fraction.

Then you multiply this by the amount.

> Remember that 'of' in mathematics usually means 'multiply by' (you met this when finding a fraction of an amount in Chapter 4).

Worked example 1

Work out 35% of 1800.

$$\frac{35}{100} \times 1800 = 630$$

> 35% is $\frac{35}{100}$ as a fraction

So, 35% of 1800 is 630.

Worked example 2

Fatou sees a toy costing $80 in a shop.

She saves 3% by buying it on the internet.

How much does she save?

$$\frac{3}{100} \times 80 = 2.40$$

She saves $2.40

Worked example 3

At a football match there are 36 000 spectators.

85% of the spectators are adults.

How many adults are there?

$$\frac{85}{100} \times 36\,000 = 30\,600$$

There are 30 600 adults.

Using percentages to make a comparison

Sometimes you have to compare two percentages of different amounts, or a percentage and a fraction.

To work out which is the larger amount, you usually need to work out the actual quantity.

Worked example 4

Theo is making a flan.

He can choose to use either 100 g of Edam cheese or 90 g of Cheddar cheese.

Edam cheese contains 25% fat and Cheddar cheese contains $\frac{1}{3}$ fat.

Theo wants to use the cheese that will add the smallest amount of fat to his flan.

Which cheese should he use?

Edam cheese

25% of 100 g is $\frac{25}{100} \times 100 = 25$ g fat

Cheddar cheese

$\frac{1}{3}$ of 90 g is $\frac{1}{3} \times 90 = 30$ g fat

The Edam cheese contains less fat, so Theo should choose Edam cheese for the flan.

Exercise 17.2

You must not use a calculator in this exercise.

1 Find the value of:

 a 50% of $30   **b** 30% of $60   **c** 40% of $50

 d 60% of $80   **e** 25% of $72   **f** 10% of $80

2 Work these out:

 a 35% of 400 cm   **b** 25% of 12 kg   **c** 15% of 240 g

 d 5% of 20 litres   **e** 80% of 320 m   **f** 75% of 1 hour

3 240 people watch a netball match.

25% of the people watching are children.

 a How many children watch the netball match?

 b How many people watching the match are not children?

4 90% of the machines in a factory are working correctly.

There are 800 machines in the factory.

How many machines are **not** working correctly?

5 A tree is 650 cm tall at the start of summer.

During summer the tree grows and its height increases by 8%.

By how many cm does the trees height increase?

6 Huan is going on holiday.

The journey is 640 km.

Her mum says they have travelled 30% of the distance.

How many km do they have left to travel?

7 Gravel loses 10% of its mass when it dries.

A load of wet gravel has a mass of 450 kg.

What will the mass of the gravel be when it is dry?

8 Two shops sell the same computer.

These are their advertisements.

Shop A   Shop B

Normal price $760 You only pay 75%

Normal price $700 You only pay 80%

Which shop sells the computer at the lower price?

9 Which is larger, $\frac{2}{3}$ of 75 metres or 80% of 60 metres?

10 A pack contains 50 seeds and costs $5. Erika sows all the seeds. 40% of the seeds grow into plants. She sells the plants for $1 each. How much profit does she make?

11 The insurance premium for Rachel's car is $960. It is reduced by 60% for a 'no claims discount'. Rachel pays this in 12 equal monthly payments. What is each monthly payment?

12 A pack contains 240 seeds. Sanjay sows all the seeds. 60% germinate. 75% of those that germinate grow into plants. $\frac{1}{3}$ of the plants grow yellow flowers, the rest grow red flowers.

How many plants have red flowers?

> Start by finding 60% of 240. Then work out 75% of your answer to find the number that grow into plants.

Review

1 For each of the following shapes, write down:

i the fraction that is shaded

ii the percentage that is shaded.

a 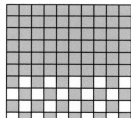 **b**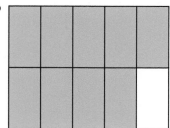

2 There are 400 music CDs in a CD rack.

120 are pop

140 are rock

80 are dance

60 are classical

a What fraction are: **i** pop **ii** dance?

b What percentage are: **i** rock **ii** classical?

3 Lucy takes tests in three subjects.

Her results are:

38 out of 50 in statistics

77% in economics

16 out of 20 in geology.

a In which subject did Lucy do best?

b What was her percentage in this subject?

4 Work these out:

    a 50% of 20 m

    b 20% of 15 kg

    c 15% of 120 minutes

    d 60% of 1200 kg.

5 In a class of 30 students 60% are female.

    a What percentage of the students are male?

    b How many of the students are male?

6 **a** A train journey has a fare of $70.

      The fare for the journey increases by 5%.

      How much is the increase?

    b A bus journey has a fare of $6.50

      Liam has a student card that gets 8% discount.

      How much discount does Liam get on this bus journey?

7 A car costs $16 500. The price is reduced by 15% in a sale.

    a By how much is the price reduced?

    b What is the sale price of the car?

8 Which is larger, $\frac{1}{3}$ of 48 or 30% of 50?

9 A holiday company improves the standard of a holiday.

To cover the costs it increases the price by 9%.

What is the new price of a holiday that originally cost $800?

18 Planning and collecting data

Chapter 6 covers processing data
Chapter 10 covers presenting data

Learning outcomes

- Decide which data would be relevant to an enquiry.
- Collect and organise data.
- Design and use a data collection sheet.
- Design and use a questionnaire for a simple survey.
- Construct and use frequency tables to gather discrete data, grouped where appropriate in equal class intervals.

Useful data

There should always be a reason for collecting data.

Your teachers need information, such as the number of students in each of their classes, to work out the number of desks and chairs they need.

Your parents need to know the family income to plan spending on food, bills and clothing.

The branch of mathematics dealing with collecting, analysing and interpreting data is called statistics. The word 'statistics' is used when summarising the data for interpretation. You have already used some statistics when you learned about the mode, mean, median and range.

Scientists, industry and insurance companies often use statistics.

Governments need data to plan how many hospitals, roads and schools to build.

Sports clubs collect data to find out about team performance and individual performance.

18.1 Data collection sheets

Deciding on which data are needed

There should always be a reason for collecting data. This reason will help you decide what data should be collected. You may want to know what food a group of people prefer, or how busy a road is.

Some data may be collected by observation, experiment or by using a questionnaire. In all of these cases the type of data to be collected needs to be decided first.

Worked example 1

Julia wants to find out whether boys like rock music more than girls.
What data must she collect?

Julia must find out if the people she asks are male or female. She must also ask them whether they like rock music or not.

Worked example 2

Felipe wants to find out whether older students are more likely to walk to school than younger students. What data must he collect?

Felipe must ask the students for their age. He must also ask if they walk to school or not.

Data collection sheets

Once you have decided on what data needs to be collected you need to decide how to collect and record it. One way is to use a **data collection sheet**.

A data collection sheet is a table with headings for the different items you want to record.

You can use a tally mark (/) to record each item in the correct column of the table.

When you reach 5 items in a column, the fifth tally goes across the other 4 to make a 'five-bar-gate'.

$$\underset{3}{///} \quad \underset{5}{\cancel{||||}} \quad \underset{7}{\cancel{||||}\ //}$$

Grouping the tallies in 5s makes it easier to add them up when you have collected all the data.

Worked example 3

Harry wants to find out what colour car is most popular with his teachers.

He records the colour of the cars in the school car park.

Design a data collection sheet that he can use to do this. Show how Harry would add a tally mark in the correct part of the table for each car.

Colour	Red	Blue	Black	White	Green	Grey
Tally	///	$\cancel{||||}$ /	$\cancel{||||}$ ///	////	//	/

Worked example 4

A head teacher wants to find out how many students in each class are wearing the correct school uniform. The school uniform is tie, white shirt, black trousers and black shoes. Design a data collection sheet that could be used by class teachers.

Show how the data collection sheet could be filled in by the teachers.

The following data collection sheet would be suitable.

Student	Tie	White shirt	Black trousers	Black shoes
A	✓	✓	✓	✗
B	✓	✗	✓	✗
C	✗	✗	✗	✓

A tally chart, like the one in Worked example 3 would only record the total number of school uniform items that were not being worn. It would not show how many items each student was wearing.

Exercise 18.1

1 Clara wants to find out whether boys or girls watch more television.

What data should Clara collect?

2 Frank thinks that holidays in hotter countries are more expensive than holidays in colder countries. What data does Frank need to collect to find out if he is correct?

3 Kerry says 'more girls play sport than boys.' Lee says that Kerry is not correct as more boys play soccer than girls. Why does Lee's statement not disagree with Kerry's statement?

4 Hassan says 'the most popular colour for front doors in my street is red.'

 a What data should Hassan collect to find out whether his statement is correct?

 b Design a data collection sheet he could use.

5 A mathematics teacher wants to know if her students have the correct equipment for her lessons. Design a data sheet she can use.

6 Manju wants to find out the most popular way students get to school. Design a data collection sheet that could be used to collect the information needed.

7 Design a data collection sheet that can be used to investigate the number of people in cars as they drive past a school.

8 Design data collection sheets that can be used to find out the following:

 a the most popular food for students in your class

 b the favourite type of music of students in your class

 c the favourite type of television programme for students in your class.

9 Marcy wants to find out which of her hens lays most eggs. Her hens are called Leggy, Peggy, Chicken, Licken and Shelley. Design a data collection sheet she can use.

10 Design a data collection sheet that can be used to compare the length of words from two different sources, for example, books, magazines, newspapers.

Collect the data from the two sources you have chosen.

Compare the data using diagrams, average values and the range.

For each source you will need to know the number of letters in each word.

18.2 Questionnaire design

When you obtain data from people it is important that each person is asked exactly the same question. A series of questions is often called a questionnaire.

Surveys can use questionnaires to find out information. Sometimes the person answering the question fills in the questionnaire. Sometimes the questions are read out and filled in by the person carrying out the survey.

Questions may be open or closed.

Open questions have no suggested answers. They allow the person answering to give any answer they wish. You may get answers you were not expecting.

Sometimes this is the only way to ask a question. This is usually when there are too many possible answers to list.

Closed questions give a list of suggested answers to choose from.

It is easier to compare questionnaires that use closed questions.

Questions should be:

- easy to understand
- as short as possible.

Questions should not:

- be leading (they should not be questions that use the words 'do you agree that …')
- ask for personal information.

Option boxes or tick boxes are used for the answers to closed questions.

The answers to closed questions should include all possible answers.

Worked example 1

A librarian wants to find out how many books people read.

He wants to ask one of these questions in a questionnaire.

1   How many books did you read last week?   Answer _____

Or

2   Tick the box for the number of books you read last week.

☐ 1 to 3

☐ 3 to 5

☐ 7 or more

a   What problems could there be with these questions?

b   Write an improved question.

---

a Question 1 is an open question.

Some people may answer with fractions of a book.

Some people may answer by using words, such as 'not many', or 'more than 1', 'don't know', or even 'I'm not saying.'

Question 2 is a closed question. It is easier to answer, so people are more likely to answer it.

The options for the answers are badly chosen. If a person did not read any books or read 6 books there is nowhere for them to answer. If a person read 3 books they would not know which box to tick, as 3 is in two options.

b Tick the box for the number of books you read last week.

☐ 0

☐ 1

☐ 2

☐ 3

☐ 4 or more

Worked example 2

A café owner wants to find out his customers' incomes.

He decides to ask the following question.

How much do you earn? _____

a What problems could there be with this question?

b Write an improved question.

a The question is open so people can answer in any way they wish. Answers could include 'quite a lot', 'not much' or 'mind your own business.'

Some people may not want to share this personal information with others. One way to help to avoid this would be to use grouped answers so that exact amounts do not need to be given.

There is no time scale. Is the question asking for pay per hour, day, week, month or year?

b Tick a box to show the amount you were you paid last week.

☐ $0 to $99.99

☐ $100 to $199.99

☐ $200 to $299.99

☐ $300 to $399.99

☐ $400 or over

Exercise 18.2

1 For each of the following questions:

    a write down any problems you can see with the question

    b write an improved question.

    i How old are you? Answer _____

    ii Everyone likes films. Which film is your favourite? Answer _____

    iii Do you travel by bus? Tick a box.

        ☐ Not very often

        ☐ Sometimes

        ☐ Always

    iv What is your favourite football team? Tick a box.

        ☐ Real Madrid

        ☐ Barcelona

    v What sport do you like? Tick one box only.

        ☐ Soccer

        ☐ Cricket

        ☐ Netball

        ☐ Lacrosse

    vi How much television do you watch? Answer _____

    vii Everyone likes watching soccer on TV. Do you agree? Tick a box.

        ☐ Agree strongly

        ☐ Agree

        ☐ Not sure

    viii How many films do you watch each month? Tick a box.

        ☐ 1 to 5

        ☐ 5 to 10

        ☐ more than 12

    ix How often do you eat in a restaurant? Tick a box.

        ☐ Not often

        ☐ Quite a lot

        ☐ All the time

2 Write a short questionnaire to find out where people went on holiday in the last year.
Find out about where they went, what type of accommodation they stayed in and what they
thought was good or bad about the accommodation and food.

3 A company is thinking of opening a new leisure centre.

The company wants to carry out a survey to find out how people would use it.

Write two or three questions that could be used in a questionnaire.

18.3 Frequency tables

Frequency tables for ungrouped data

Frequency tables can be used to organise data that is in a list. They can also be used to help with data collection.

Worked example 1

Lola counted the number of people in cars as they drove past her school. She recorded her results.

2, 3, 1, 3, 2, 4, 2, 3, 1, 1,

2, 3, 2, 4, 2, 5, 4, 4, 2, 1,

2, 4, 3, 6, 3, 2, 3, 2, 5, 2,

1, 1, 3, 1, 1, 3, 2, 3, 3, 1

Put these results in a frequency table. Use a table with three columns.

* Label the left hand column 'Number of people in car'. Start at 1 (the smallest value in the list) and go up to 6 (the largest value in the list).

* Label the middle column 'Tally'.

* Label the third column 'Frequency'.

Go through the list one number at a time and fill in the tally column with tally marks (/) as you go.

Remember to use a 'five-bar-gate' (卌) for the fifth tally. This will help when you fill in the frequency column after you have done all the tallies.

Number of people in car	Tally	Frequency
1	卌 ////	9
2	卌 卌 //	12
3	卌 卌 /	11
4	卌	5
5	//	2
6	/	1

Frequency tables for grouped data

When there are a lot of different values, such as amounts of money, it is helpful to group the data.

Worked example 2

John collected data about how much his friends were paid daily for their part-time jobs.

His results were:

$9.50, $7.25, $28.90, $8.40, $36.50, $38, $29, $62, $80, $120,

$18.90, $48, $68, $52.50, $96, $112, $49, $86, $92, $32.50,

$9, $77.25, $26.50, $78, $38.50, $35, $27, $42, $30, $10,

$12, $48.50, $62, $42.50, $66, $122, $44, $81, $96, $37.50

Put the data into a grouped frequency table.

Part-time job pay ($)	Tally	Frequency
0 to 19.99	⅂⅂⅂⅂ //	7
20 to 39.99	⅂⅂⅂⅂ ⅂⅂⅂⅂ /	11
40 to 59.99	⅂⅂⅂⅂ //	7
60 to 79.99	⅂⅂⅂⅂ /	6
80 to 99.99	⅂⅂⅂⅂ /	6
100 to 119.99	/	1
120 to 139.99	//	2

A disadvantage of grouping data like this is that you no longer know what the original values were. The big advantage is that you get a good idea of the distribution of the data.

Exercise 18.3

1 The list below shows the number of people living in each house in one street.

3	2	2	3	6	3	2	6	1	4
2	1	4	5	4	1	3	3	4	3
6	2	1	4	1	4	2	4	4	4
6	6	1	5	2	2	2	2	2	5
1	1	4	1	4	2	5	6	1	5

Copy and complete the frequency table to show this information.

Number of people in house	Tally	Frequency
1		
2		
3		
4		
5		
6		

2 Seth wants to find out which colour is the most popular colour for cars.

He records the colours of the cars in the school car park.

Red	Blue	Green	Blue	Yellow	Green	Red	Green	Black	Grey
Green	Red	Red	Silver	Black	Red	Red	Blue	Red	Red
Blue	Grey	White	Blue	White	Blue	Black	Red	Blue	Black
White	Blue	White	Red	Yellow	Blue	Yellow	Red	Red	Green

a Put this information in a frequency table. Use the headings shown.

Colour of car	Tally	Frequency

b Which colour is the most popular colour?

c Was it easier to use the list or the table to answer part **b**? Give a reason for your answer.

3 A worker records the number of people in each car for 100 cars leaving a supermarket car park.

The results are:

2	2	1	5	1	1	2	4	5	3
1	1	2	4	1	2	5	5	2	4
5	2	3	4	3	4	2	1	4	2
3	3	4	5	5	1	3	3	2	3
1	1	1	3	1	4	3	4	5	3
2	3	3	2	5	4	5	5	2	4
3	3	5	3	2	3	1	2	3	2
2	5	3	1	5	1	2	4	5	5
2	2	3	1	1	2	2	4	4	4
5	5	4	5	1	3	5	3	2	2

Copy and complete the frequency table to show this information.

Number of people in car	Tally	Frequency
1		
2		
3		
4		
5		

4 A score keeper records the number of runs scored by a batsman in 50 cricket matches.

66	67	4	50	84	3	49	35	100	5
54	69	80	2	5	46	45	65	22	35
32	22	73	10	19	92	7	53	1	51
27	53	64	66	88	49	1	86	43	56
10	76	38	73	75	3	42	87	44	85

Copy and complete the grouped frequency table to show this information.

Number of runs	Tally	Frequency
0 to 19		
20 to 39		
40 to 59		
60 to 79		
80 to 99		
100 to 119		

5 Some members of a youth club are asked how much money they have with them. The amounts are:

$0.20	$1.20	$5	$0.80	$6	$10	$7.50	$3.20	$3.30	$8.95
$1.40	$2.20	$4	$4.80	$7	$9.10	$6.50	$2.40	$2.35	$3.90
$0.60	$0.90	$7.10	$3.60	$2.20	$11	$1.50	$7.20	$9.30	$2.55

Make a copy of each grouped frequency table. Complete the tables to show the information.

a i

Amount ($)	Tally	Frequency
0.00 to 0.49		
0.50 to 0.99		
1.00 to 1.49		
1.50 to 1.99		
2.00 to 2.49		
2.50 to 2.99		
3.00 to 3.49		
3.50 to 3.99		
4.00 to 4.49		
4.50 to 4.99		
5.00 to 5.49		
5.50 to 5.99		
6.00 to 6.49		
6.50 to 6.99		
7.00 to 7.49		
7.50 to 7.99		
8.00 to 8.49		
8.50 to 8.99		
9.00 to 9.49		
9.50 to 9.99		
10.00 to 10.49		
10.50 to 10.99		
11.00 to 11.49		
11.50 to 11.99		

ii

Amount ($)	Tally	Frequency
0.00 to 0.99		
1.00 to 1.99		
2.00 to 2.99		
3.00 to 3.99		
4.00 to 4.99		
5.00 to 5.99		
6.00 to 6.99		
7.00 to 7.99		
8.00 to 8.99		
9.00 to 9.99		
10.00 to 10.99		
11.00 to 11.99		

iii

Amount ($)	Tally	Frequency
0.00 to 1.99		
2.00 to 3.99		
4.00 to 5.99		
6.00 to 7.99		
8.00 to 9.99		
10.00 to 11.99		

iv

Amount ($)	Tally	Frequency
0.00 to 4.99		
5.00 to 9.99		
10.00 to 14.99		

b Why is it not a good idea to put the data into an ungrouped frequency table?

c Which one of the tables do you think is most useful? Give a reason for your answer.

Review

1 Lydia wants to find out whether 2-door, 3-door, 4-door or 5-door cars are most popular.

She wants to collect the data from the road by her school.

Design a data collection sheet that she could use.

2 Shivi wants to find out how long it takes students to get to school.

The question she asks is:

'How long does it take you to get to school?'

   a Give a reason why this question will not give Shivi the information she wants.

   b Design a data collection sheet that will allow her to collect the information she wants.

3 Andrew wants to find out whether people buy music from music shops or the internet.

Design a question that could be used in a questionnaire to collect this information.

4 A student records the number of children in each of the families of the students in his class.

His results are:

2	1	1	1	1	3	2	3	3	1
1	2	1	3	3	2	1	2	4	1
2	2	5	4	2	4	3	3	1	2
3	4	4	1	4	2	2	4	3	2

   a Record these in a frequency table.

   b Which is the most common number of children?

5 A large class of students get the following marks in a test.

51	29	40	52	31	35	58	76	59	41
66	63	42	68	72	32	44	47	79	66
21	51	55	39	68	75	65	69	42	56
65	33	21	32	53	50	48	47	39	62

Copy and complete the grouped frequency table to show this information.

Mark	Tally	Frequency
21 to 30		
31 to 40		
41 to 50		
51 to 60		
61 to 70		
71 to 80		

19 Ratio and proportion

Learning outcomes

- Use ratio notation, simplify ratios and divide a quantity into two parts in a given ratio.
- Recognise the relationship between ratio and proportion.
- Use direct proportion in context; solve simple problems involving ratio and direct proportion.

Mixing it

Many people use ratio in their everyday lives.

Builders mix sand, gravel and cement in different ratios.

For example, four times as much sand as cement makes a good mortar for bricklaying.

Six times as much gravel as cement makes a strong concrete to use as a floor.

For maximum strength the mass of the water in the mix should be four times the mass of the cement.

19.1 Ratio

Working with ratio

This is a pattern of tiles along a wall.

There are two red tiles and four blue tiles.

The ratio of red tiles to blue tiles is written as 2 : 4. This is read as 2 **to** 4.

Each red tile has a blue tile on either side of it.

The ratio can be written in a simpler form as 1 : 2

Here is a longer piece of the pattern.

The ratio of red tiles to blue tiles here is 6 : 12

The numbers 6 and 12 have a highest common factor of 6. Divide both numbers by 6 to simplify the ratio.

The simplified ratio is 1 : 2

> Simplifying ratios is similar to cancelling fractions, which you met in Chapter 4.

Worked example 1

Here is a row of small flags.

a Write down the ratio of green flags to blue flags.

b Write the ratio in its simplest form.

a There are 9 green flags and 6 blue flags. The ratio of green flags to blue flags is 9 : 6

b The numbers 9 and 6 have a highest common factor of 3. Divide both numbers by 3.

The ratio in its simplest form is 3 : 2

> Always check at the end of a question that the ratio cannot be simplified again.

Worked example 2

Here is part of a recipe for making shortbread.

600 g plain flour

450 g sugar

200 g butter

Work out these ratios. Give your answers in their simplest form.

a The ratio of flour to butter.

b The ratio of sugar to flour.

a The ratio of flour to butter is 600 : 200

The highest common factor is 200. Divide by 200.

The simplified ratio is 3 : 1

b The ratio of sugar to flour is 450 : 600

The highest common factor is 150. Divide by 150.

The simplest ratio is 3 : 4

> It is important that you write the numbers in the correct order. The ratio of sugar to flour is 3 : 4, not 4 : 3

Exercise 19.1

1 In each of these patterns of flags write down the ratio of red flags to blue flags.

a

b

c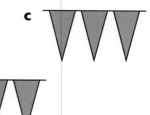

d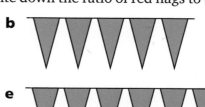

e

2 For each of these patterns write the ratio of yellow squares to green squares in its simplest form.

a

b

c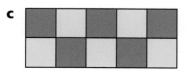

d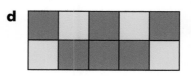

e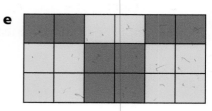

3 Here is a row of flags.

a Write down the ratio of black flags to yellow flags in its simplest form.

b Write down the ratio of yellow flags to black flags in its simplest form.

4 Simplify each of these ratios.

   a 6 : 4   **b** 15 : 20   **c** 12 : 10   **d** 8 : 16   **e** 24 : 12   **f** 35 : 15

   g 18 : 27   **h** 70 : 50   **i** 2 : 32   **j** 21 : 12   **k** 16 : 16   **l** 15 : 33

5 a Simplify the following ratios.

     i 16 : 12   **ii** 15 : 20   **iii** 8 : 6   **iv** 5 : 8   **v** 8 : 5   **vi** 36 : 27

   b Write down the three ratios that are equivalent.

6 In class 12L there are 18 girls and 15 boys.

    a What is the ratio of girls to boys?

    b What is the ratio of boys to girls?

    Give both answers in their simplest form.

7 Here is part of a recipe for cheese scones.

        240 g flour

        A pinch of salt

        60 g butter

        30 g cheese

        150 ml milk

    a Write down the ratio of flour to butter. Simplify your answer.

    b Write down the ratio of cheese to butter. Simplify your answer.

    c Juan has only 20 g of cheese and wants to use all of it to make scones.

       i How much butter should he use?

      ii How much flour should he use?

> Divide everything by 3 to find how many scones can be made with 10 g cheese.

    d Samantha says that the ratio of cheese to milk is 1 : 5

      Is she correct?

      Explain your answer.

> Remember that the units must be the same when writing ratios.

8 Simplify each of these ratios.

    a $\frac{3}{4} : \frac{5}{8}$ **b** $\frac{3}{10} : \frac{4}{5}$ **c** $\frac{1}{2} : \frac{5}{8}$ **d** $\frac{2}{3} : \frac{7}{9}$ **e** $\frac{3}{4} : \frac{5}{12}$

> Change $\frac{3}{4}$ into $\frac{6}{8}$. The ratio is then $\frac{6}{8} : \frac{5}{8}$

19.2 Dividing in a given ratio

You can share things out so that they are divided in a given ratio.

Polly and Alice want to share 12 sweets in the ratio 3 : 1

For every 3 sweets that Polly takes, Alice takes 1.

Polly takes 3 times as many sweets as Alice.

This diagram shows how they can share out the sweets.

First you add up how many portions there are in total. $3 + 1 = 4$

The girls share the sweets out into 4 equal portions.

Polly has 3 portions and Alice has 1 portion.

Polly has 9 sweets and Alice has 3 sweets.

This is what you do to share a quantity in a given ratio:

+ **Add to find the total number of portions.**

÷ **Divide to find the size of each portion.**

× **Multiply to find the size of each share.**

Worked example 1

Anna and Rosalind share $120 in the ratio $1 : 3$

Work out how much each will receive.

$1 + 3 = 4$	**add** to find the total number of portions
$120 \div 4 = 30$	**divide** to find the size of each portion
Anna $= 30 \times 1 = 30$	**multiply** to find the size of each share
Rosalind $= 30 \times 3 = 90$	

Anna receives $30 and Rosalind receives $90.

> You can check at the end that the two numbers add up to the correct total. $30 + 90 = 120$

Worked example 2

Antonio and Pablo want to buy a car that costs $2400.

They agree to share the price in the ratio $3 : 2$

Work out how much each will pay.

$3 + 2 = 5$

$2400 \div 5 = 480$

$480 \times 3 = 1440, 480 \times 2 = 960$

Antonio will pay $1440 and Pablo will pay $960. •———(Check that $1440 + 960 = 2400$

Worked example 3

Jasper and Matt share some money in the ratio $5 : 2$

Jasper receives $300.

a How much does Matt receive?

b Work out how much money there is altogether.

a Jasper receives 5 portions.

$300 \div 5 = 60$ each portion is worth $60

Matt receives 2 portions. $2 \times 60 = 120$

Matt receives $120.

> Check that $300 : 120$ simplifies to $5 : 2$

b The total amount of money is $300 + $120 = $420

Exercise 19.2

1 Saeed and Khalil share 35 plants in the ratio $3 : 4$. Who takes more, Saeed or Khalil?

2 Divide these amounts in the given ratios.

   a $80 in the ratio $2 : 3$ **b** $200 in the ratio $4 : 1$ **c** $140 in the ratio $5 : 2$

   d 63 kg in the ratio $2 : 7$ **e** 35 sweets in the ratio $6 : 1$ **f** 65 plants in the ratio $4 : 9$

   g $99 in the ratio $10 : 1$ **h** $500 in the ratio $3 : 7$ **i** $24 in the ratio $3 : 2$

3 Phansit and Mutaya share the rent of their flat in the ratio $4 : 3$. The rent is $420 each month.

Work out how much they each pay each month.

4 On a school visit to a museum there are 84 students altogether.

The ratio of boys to girls is $3 : 4$

Work out how many boys there are on the visit.

5 Tom has 45 roses in his garden. The ratio of red roses to white roses is $4 : 1$

Work out how many of each colour he has.

6 Superfilms and Moviestars are two companies who are making a film.

They agree to share the cost in the ratio $5 : 3$. The film costs $720 000 to make.

Work out how much each company must pay.

7 Jackie and Jon buy a car together. Jackie pays $1200 and Jon pays $2100.

   a Work out the ratio of their payments. Give the answer in its simplest form.

Five years later they sell the car for $550. They agree to share the money in the same ratio.

   b Work out how much they each receive when they sell the car.

8 Arin and Tamoy share some money in the ratio $1 : 3$

What fraction of the money does each receive?

9 A bag contains some red beads and some blue beads.

The ratio of red beads to blue beads is 5 : 3. There are 75 red beads in the bag.

Work out the total number of beads in the bag.

10 At a football match the ratio of home supporters to away supporters is 4 : 1

There are 12 800 home supporters at the match.

Work out how many away supporters there are.

11 Hal and Alma want to share 2 litres of milk in the ratio 7 : 3

Work out how much milk each receives. Give your answer in millilitres.

12 Andrew and Richard share a sum of money in the ratio 3 : 5

Richard receives $320 more than Andrew.

How much does each of them receive?

> Richard receives two portions more than Andrew. So two portions = $320.

13 $\frac{5}{8}$ of the children in a class are boys.

What is the ratio of boys to girls?

14 $\frac{4}{7}$ of the roses in my garden are red. The rest are white.

Work out the ratio of red roses to white roses.

> If $\frac{5}{8}$ are boys then $\frac{3}{8}$ are girls.
>
> Simplify this ratio.

15 $\frac{3}{5}$ of the audience in a cinema are female.

Work out ratio of males to females in the cinema.

19.3 Proportion

Proportion

Sometimes you need to work out if ratios between quantities are the same.

When the ratios are the same, the quantities are in proportion.

This example shows how to do this.

Worked example 1

Antonio has mixed 3 litres of red paint with 2 litres of white paint.

Fernando has mixed 750 millilitres of red paint with 500 millilitres of white paint.

Are the mixtures in proportion?

Look at the ratio of red : white for each of them.

Antonio	3 l : 2 l	Fernando	750 ml : 500 ml	
	= 3 : 2		= 750 : 500	divide by 250 to simplify the ratio
			= 3 : 2	

They have both mixed the colours in the ratio 3 : 2 so the mixtures are in proportion.

Direct proportion

When two quantities are always in the same ratio they are in **direct proportion**.

You can multiply or divide both quantities by the same number, and the ratio stays the same.

For example, mixing sand and cement in the ratio 4 : 1 makes mortar.

How much sand is needed to mix with
25 kg cement?

Using multiplication:

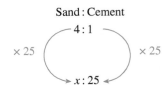

Sand : Cement

So $x = 4 \times 25 = 100$ kg

Both quantities have been multiplied by the same number.

The ratio has remained the same.

The ratio 4 : 1 is equivalent to the ratio 100 : 25

Worked example 2

Victoria is paid $5.30 per hour at work.

How much is she paid for working 4 hours?

For 1 hour she is paid $5.30.
For 4 hours she is paid $4 \times 5.30 = 21.20$
She is paid $21.20 for four hours' work.

Worked example 3

Mimi buys 6 pens for $4.20.

Work out the cost of: **a** 12 pens   **b** 3 pens   **c** 15 pens.

a 12 pens cost twice as much as 6 pens.

$$2 \times 4.20 = 8.40$$

She pays $8.40 for 12 pens.

b 3 pens cost half as much as 6 pens.

$$4.20 \div 2 = 2.10$$

She pays £2.10 for 3 pens.

c 15 pens cost 5 times as much as 3 pens.

$$2.10 \times 5 = 10.50$$

She pays $10.50 for 15 pens.

Exercise 19.3

1 Amiira mixes 500 g flour with 200 g sugar in a cake mix.

Hassan mixes 360 g flour with 120 g sugar in his cake mix.

   a Work out the ratio of flour to sugar for Amiira's cake. Simplify the answer.

   b Work out the ratio of flour to sugar for Hassan's cake. Simplify the answer.

   c Have they mixed flour and sugar in the same proportion? Explain your answer.

2 Anderson is paid $6.50 for one hour's work. How much is he paid if he works for:

   a 4 hours   **b** 11 hours?

3 Bottles of juice cost 89 cents each. How much would the following cost?

   a 5 bottles of juice   **b** 12 bottles of juice   **c** 25 bottles of juice

4 12 exercise books have a total mass of 504 grams.

Work out the mass of:

   a 6 exercise books   **b** 3 exercise books   **c** 24 exercise books.

5 Tomasz is paid $23.36 for 4 hours' work.

How much is he paid for 12 hours' work?

6 Curtain material costs $44.95 for 5 metres.

   a Work out the cost of 10 metres of curtain material.

   b How much would 1 metre of material cost?

7 Five cricket balls have a total mass of 800 grams.

   a Work out the mass of 1 cricket ball.

   b Work out the mass of 8 cricket balls.

8 This is a recipe for banoffee pie. It makes enough pie for 4 people.

   a How many bananas are needed to make banoffee pie for 8 people?

   b What mass of butter is needed to make banoffee pie for 2 people?

   c Write out the recipe to make banoffee pie for 12 people.

Banoffee Pie
400 g crushed biscuits
200 g butter
2 tins condensed milk
0.5 l double cream
2 bananas

9 The instructions on a bottle of fruit squash say:

'Mix one part squash with six parts water.'

   a Write this as a ratio.

   b If you use 80 ml of squash, how much water is needed?

10 A French tourist changes some euros into US dollars at a bank.

She receives $1.31 for each euro.

How much does she get for changing 160 euros?

11 Roberta pays $19.50 for 6 pizzas.

    a Work out the cost of 3 pizzas.   **b** Find the cost of 21 pizzas.

12 32 children travel by train to the zoo.

They each pay the same amount.

The total train fare for the 32 children is $768.   Work out the cost per person first.

The following day 36 children decide to go to the zoo.

Work out the total train fare for 36 children if the fare for each child is still the same.

13 Three tourists from Argentina arrive in the United States for a holiday.

They exchange some Argentine pesos into US dollars.

They all exchange their money at the same exchange rate.

Alicia changes 500 pesos and receives $95.

    a Juan changes 800 pesos.

       Work out how many dollars he receives.   Use Alicia's figures to work out the exchange rate.

    b Cristina receives $125.40 for her pesos.

       Work out how many pesos she changes.

Review

1 Here is a set of flags.

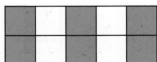

    a Write down the ratio of yellow flags to blue flags. Simplify your answer.

    b Write down the ratio of blue flags to yellow flags. Simplify your answer.

2 For each of these tiling patterns give the ratio of red tiles to white tiles in its simplest form.

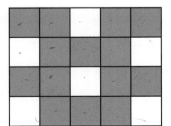

3 Write these ratios in their simplest form.

    a $12:10$ **b** $8:14$ **c** $21:28$ **d** $32:72$ **e** $125:30$

4 Simplify these ratios:

    a 3 weeks : 6 days   **b** $600\,\text{ml}:4\,\text{l}$ **c** $750\,\text{g}:2.4\,\text{kg}$

5 For each of the following paper sizes work out the ratio of the length to the width.

In each case give your answer as $1:n$

  a A4 paper, which is 297 mm by 210 mm.

  b A3 paper, which is 420 mm by 297 mm.

  c A1 paper , which is 841 mm by 594 mm.

  d What do you notice?

  e Find out some more paper sizes and see if this always works.

6 Divide the following quantities in the given ratios.

  a 240 sweets in the ratio $3:1$

  b 175 biscuits in the ratio $1:4$

  c $280 in the ratio $2:5$

  d $1452 in the ratio $7:4$

7 There are 24 310 people at a football match. The ratio of males to females is $10:1$

How many males are at the match?

8 Each year Tara and Ben's aunt gives them some money.

They share it in the ratio of their ages.

  a She gives them $500 when Tara is 4 years old and Ben is 6 years old.

    How much do they each receive?

  b The following year she gives them $540.

    How much do they receive this year?

> Remember they are now a year older.

9 Jeremy pays $11.20 for 4 packs of file paper.

  a Find the cost of 8 packs of paper.

  b Find the cost of 2 packs of paper.

10 A coach company charges customers 45 cents per mile.

Robert travels 14 miles to work.

What is his coach fare? Give your answer in dollars.

11 This is a recipe for shortcrust pastry. It makes 12 pies.

  a What is the ratio of flour to butter?

  b How much butter is needed to make 24 pies?

  c Write down the quantities needed to make 6 pies.

  d Write down the quantities needed for 30 pies.

> **Shortcrust pastry**
> 150 g plain flour
> A pinch of salt
> 75 g butter
> 45 ml water
>
> *Makes 12 pies*
>

12 Jim is 6 years older than Arthur. The ratio of their ages is $4:7$

  a How old are they?

  b What will be the ratio of their ages next year?

Time and rates of change

Measuring time

People have tried to measure time since the earliest civilisations.

Some of our units of time are based on a number system used in Babylon over 4000 years ago.

Many different methods have been used to measure time. One early method was a candle marked in hours. Another method was the hourglass, like the one shown here, which takes an hour for the sand to go through.

These were very inaccurate methods compared to the modern, digital clocks that we are all familiar with.

20.1 Time

Units of time

These are the basic units of time:

1 minute = 60 seconds; 1 hour = 60 minutes; 1 day = 24 hours; 1 week = 7 days

For longer periods there are months and years. Months vary in length from 28 days to 31 days. A year is normally 365 days but a leap year has 366 days. In this chapter you will only use weeks, days, hours, minutes and seconds.

12- and 24-hour clock

Traditional clocks, like the one shown on the next page, repeat every 12 hours. Each time occurs twice per day. This is the 12-hour clock system.

This clock shows 9.15. You do not know if it is in the morning or the evening.

You have to state am or pm to show whether it is before midday or after midday.

9.15am means quarter past nine in the morning.

9.15pm means quarter past nine in the evening.

This clock has a digital display. It repeats every 24 hours, starting at midnight. This is the **24-hour clock** system.

08 10 is ten minutes past eight in the morning.

20 10, shown here, is ten minutes past eight in the evening.

> You sometimes see this written 20.10 or 2010. 24-hour times always have four digits and never have am or pm.

During the morning both systems look very similar. You need to make sure there are four digits in the 24-hour system and am in the 12-hour system.

Examples: 8.45am is the same as 08 45

11.17am is the same as 11 17

03 12 is the same as 3.12am

10 08 is the same as 10.08am

After 1pm the hour digits differ by 12 and after midday the 12-hour clock has pm.

Examples: 7.30pm is the same as 19 30

2.35pm is the same as 14 35

12.54pm is the same as 12 54

23 55 is the same as 11.55pm

18 00 is the same as 6.00pm (or just 6pm)

Worked example

A train leaves a station at 07 45 and arrives at its destination at 15 34 on the same day.

a Write these times in the 12-hour clock.

b Find the time taken for the journey.

a 07 45 = 7.45am   it is before midday so am is needed

15 34 = 3.34pm   it is after midday so pm is needed

b There are two methods for doing this.

Method 1 Use counting on to find the length of the journey.

This diagram may help.

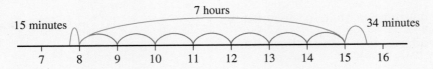

From 07 45 to 08 00 is 15 minutes.

From 08 00 to 15 00 is 7 hours.

From 15 00 to 15 34 is 34 minutes.

Total journey time is 15 minutes + 7 hours + 34 minutes = 7 hours 49 minutes.

Method 2 Use subtraction to find the length of the journey.

> Take great care as you have to remember that there are 60 minutes in each hour.

Write the subtraction with a clear gap between the hours and the minutes.

$$\begin{array}{r} 15\ \ 34 \\ -07\ \ 45 \\ \hline \end{array}$$

For this method you must write the times using the 24-hour clock.

As 45 is greater than 34, you need to change an hour into minutes.

Reduce the 15 hours by 1 to 14 and increase 34 minutes by 60 to 94.

$$\begin{array}{r} {\scriptstyle 4\ \ 94} \\ 15\ \ 34 \\ -07\ \ 45 \\ \hline \end{array}$$

This looks similar to normal subtraction but the minutes are increasing by 60, not 10.

You can now subtract the minutes and hours separately.

$$\begin{array}{r} {\scriptstyle 4\ \ 94} \\ 15\ \ 34 \\ -07\ \ 45 \\ \hline 49 \end{array} \qquad \begin{array}{r} {\scriptstyle 4\ \ 94} \\ 15\ \ 34 \\ -07\ \ 45 \\ \hline 07\ \ 49 \end{array}$$

$$94 - 45 = 49 \text{ minutes} \qquad 14 - 7 = 7 \text{ hours}$$

You now have an answer of 7 hours and 49 minutes.

Exercise 20.1

1 Work out how many:

    a minutes in 5 hours   **b** seconds in 4 minutes   **c** days in 4 weeks

    d seconds in an hour   **e** minutes in a day.

2 Write these times in the 12-hour clock.

    a 13 45   **b** 09 22   **c** 12 55

    d 21 43   **e** 00 30   **f** 11 15

3 Write these times in the 24-hour clock.

   a 5.15am   **b** 7.35pm   **c** 8.20am

   d 11.48pm   **e** 9pm   **f** 10.05am

4 How long it is between these times?

   a 09 15 and 11 49   **b** 11 45 and 17 20   **c** 03 23 and 15 11

5 How long is it between these times?

   a 11.30am and 3.20pm   **b** 5.15am and 9.12am   **c** 10.15am and 5.54pm

6 A train leaves London at 08 24 and arrives in Paris at 11 47. How long is the journey?

7 A bus leaves at 10 17 and takes 55 minutes for a trip. What time does it arrive?

8 A train departs from Adelaide at 7.45am and arrives in Melbourne 11 hours and 5 minutes later. At what time does the train reach Melbourne?

9 A flight from Auckland to Christchurch takes 1 hour and 22 minutes. If the plane departs from Auckland at 17 55 at what time does it land in Christchurch?

10 A cinema is showing a film that lasts 168 minutes.

   a How long is that in hours and minutes?

   b If the film starts at 19 45, what time does it finish?

11 The time in Singapore is 6 hours ahead of Paris. When it is 11 15 in Paris, what is the time in Singapore?

12 The time in Kolkata is $4\frac{1}{2}$ hours ahead of London. When it is 11.40am in London, what is the time in Kolkata?

13 Sunil goes away on holiday. He leaves home on Thursday and returns exactly two weeks later.

   How many days is he away for?

14 Juan starts work at 8.20am and finishes at 5.45pm.

   He has a 20 minute break each morning and a 15 minute break in the afternoon.

   He has 1 hour and 10 minutes break at lunchtime.

   How long does he work each day?

> Add up the times of the three breaks. Then subtract that from the total time spent at work.

15 Philippe starts work at 08 30 and finishes at 17 45.

   He has a 70 minute break at lunchtime and a 20 minute break in the afternoon.

   Sandrine also finishes work at 17 45.

   She has a 75 minute break at lunchtime but takes no other breaks from work.

   They both work the same number of hours each day.

   Work out at what time Sandrine starts work.

> Work out how long Philippe works first. Use that to find Sandrine's starting time.

16 Tracey wants to find out how long it is from 11 49 to 18 07.

She uses her calculator to work out 18.07 − 11.49 and gets the answer 6.58.

$$18.07 - 11.49 = 6.58$$

> Note that h and min are sometimes written to be short for hours and minutes.
>
> You will also see s written to be short for seconds.

She writes the answer as 6 h 58 min.

Explain what she has done wrong and work out the correct answer.

20.2 Using timetables

This is part of a timetable for a bus at a theme park.

Each column shows the times of a bus.

The first bus shown leaves the main entrance at 13 45. It reaches the restaurant at 14 23.

The second bus leaves the main entrance at 14 00. It reaches the restaurant at 14 38.

	Departures		
Main entrance	13 45	14 00	14 15
Funfair	13 53	14 08	14 23
Maze	14 02	14 17	14 34
Tea rooms	14 12	14 29	14 45
Racetrack	14 19	–	14 52
Restaurant	14 23	14 38	14 56

Worked example

Use the timetable to answer these questions.

a How long does the 13 45 bus take to travel from the main entrance to the restaurant?

b Jane wants to be at the maze by 2.30pm. What is the latest bus she can catch from the main entrance?

c There is a dash next to the racetrack in the middle column. What does this show?

a The bus leaves the entrance at 13 45. It reaches the restaurant at 14 23.

From 13 45 to 14 00 is 15 minutes.

From 14 00 to 14 23 is 23 minutes.

Altogether it takes 15 + 23 = 38 minutes.

b 2.30pm = 14 30. She needs to be there by 14 30.

The 14 15 bus reaches the maze at 14 34. This is too late.

The 14 00 bus reaches the maze at 14 17. She will be in time if she catches this bus.

c The dash shows that the bus does not stop at the racetrack. It goes directly from the tea rooms to the restaurant.

Exercise 20.2

1 This timetable shows flights from Singapore to Phuket.

Singapore	08 00	08 40	10 10	12 25	13 20	14 50
Phuket	09 46	10 26	11 56	14 11	15 06	16 36

a What time does the 10 10 flight arrive at Phuket?

b Meera lands at Phuket at 15 06. What time does she leave Singapore?

c How long does it take to fly from Singapore to Phuket?

d What is the latest flight I can catch to be in Phuket by 3pm?

2 This is part of a bus timetable.

	Citybus departure times			
Bus station	15 36	16 06	16 45	17 13
Art gallery	15 45	16 15	16 54	17 22
Shopping centre	15 51	16 21	17 00	17 28
Railway station	15 54	16 24	17 03	17 31
Sports ground	16 10	16 40	17 19	17 47
Town hall	16 17	16 47	17 26	17 54
Harbour	16 25	16 55	17 34	18 02

a What time does the first bus arrive at the harbour? Give the answer in 12-hour clock format.

b How long is the ride from the railway station to the sports ground?

c Jenna leaves the art gallery at 3.55pm. How long does she need to wait for the next bus?

d How long is the journey from the bus station to the harbour?

e The next bus leaves the bus station at 18 34. What time does it arrive at the harbour?

3 This is part of a railway timetable.

	Departure times				
Burgtown	07 00	10 20	12 45	15 32	18 40
Umbersford	07 43	11 05	13 28	16 17	–
Elland	–	–	13 55	–	–
Martinsville	08 45	12 10	14 32	17 20	20 28
Yarlstown	08 58	12 22	14 45	17 33	20 40
Exbury	10 10	13 35	16 05	18 45	21 55

a What time is the only train from Burgtown that stops at Elland station?

b How long does the 07 00 train from Burgtown take to arrive at Exbury?

c Michael leaves work in Umbersford at 3pm. How long does he need to wait for a train to Exbury?

d Hassan catches the 07 00 train from Burgtown to Martinsville. He spends 3 hours in a meeting in Martinsville.
What time is the next train he can catch to Exbury?

4 This is part of a timetable of trains between London and Manchester.

London Euston	09 40	10 00	10 20	10 40	11 00	11 20
Milton Keynes	–	–	10 50	–	–	11 50
Stafford	–	–	–	–	–	–
Stoke	–	11 25	11 48	–	12 25	12 48
Macclesfield	–	11 41	–	–	12 41	–
Crewe	11 11	–	–	12 11	–	–
Wilmslow	11 27	–	–	12 27	–	–
Stockport	11 37	11 56	12 17	12 37	12 56	13 17
Manchester Piccadilly	11 49	12 07	12 28	12 49	13 07	13 28

a What time is the first train from London Euston that stops at Stoke?

b How many of these trains stop at Macclesfield?

c How long does it take for the 10 40 train from London to arrive at Wilmslow?

d Nadine wants to leave her house in London at 9.45am to catch a train to Stoke.

She wants to spend an hour in Stoke before catching another train to Stockport.

Which trains could she catch?

5 This is part of John's timetable for trains between Washington DC and Boston.

	Departures				
Washington DC	08 40	10 20			14 02
Philadelphia	10 35				
New York	12 05				
Boston	16 36			19 21	

Use the times you can see in the first column. For example, the 08 40 train shows that it takes 1 hour and 55 minutes to go from Washington to Philadelphia.

John spills ink on his timetable and cannot read it all.

He knows that each train takes exactly the same time between stops.

Use the information that he can see to fill in the rest of the timetable.

20.3 Graphs

Robert goes for a ride on his bicycle.

He leaves home and rides 20 km in an hour.

He stops for a rest for 30 minutes.

He then rides for 2 hours to a picnic place, 50 km from his house.

He stops at the picnic place for 1 hour and then cycles back home without stopping.

The ride home takes him $2\frac{1}{2}$ hours.

You can show Robert's journey on a graph.

The horizontal axis of the graph shows the time.

The vertical axis shows the distance from a fixed point.

The graph is called a travel graph or a distance−time graph.

Here is the travel graph to show Robert's bicycle ride.

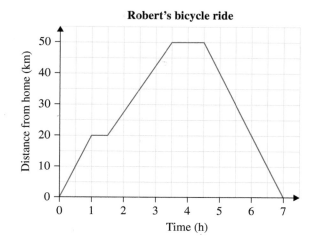

The first line on the graph shows the 20 kilometre ride for 1 hour.

The short horizontal section shows Robert's 30 minutes rest stop.

The next part of the graph ends 50 kilometres from home after a total of $3\frac{1}{2}$ hours.

The second horizontal section represents the one hour stop at the picnic place.

The final section shows his return to home in $2\frac{1}{2}$ hours.

Worked example

Use Robert's travel graph to answer these questions.

a How long is he away from home altogether?

b If he leaves home at 9am, what time does he arrive back home?

c The line for the first part of Robert's trip is steeper than the line for the second part. What does this show?

d How far does he cycle altogether?

a The line returns to the axis after 7 hours. He is away from home for 7 hours.

b 7 hours after 9am is 16 00 or 4pm.

c A steeper line shows that he is riding faster.

d He cycles 50 kilometres from home and then 50 kilometres back again. That is 100 kilometres altogether.

Exercise 20.3

1 This graph shows a train journey.

a How far does the train go in the first part of the journey?

b The journey shows a stop at a station. How many minutes does the train stop for?

c What is the total journey time from start to finish?

d How far does the train travel altogether?

e When is the train travelling fastest? Give a reason for your answer.

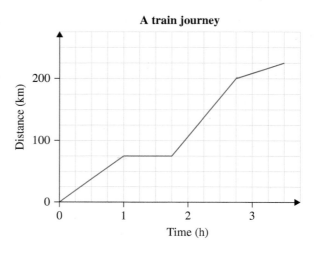

A train journey

2 Sachin goes out for a walk one day. This graph shows his journey.

a What time of day does he leave the house?

b How far does he walk altogether?

c When is he walking fastest? How can you tell?

d How many times does he stop?

e Work out how long he stops for altogether.

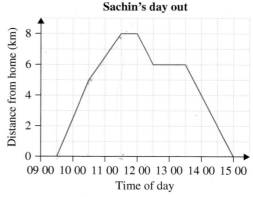

Sachin's day out

3 Alice goes for out for a drive in her car. This is her journey.

She leaves home at 10 00 and drives 120 kilometres in 2 hours.

She then stops for an hour to see her friend.

After the stop she drives 50 kilometres further from home in 30 minutes.

She then stops for another hour at some shops.

Finally she drives home without stopping.

It takes her $1\frac{1}{2}$ hours to get home.

Make a copy of these axes and draw a graph of Alice's journey.

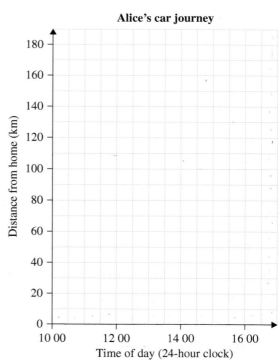

Alice's car journey

4 This graph shows the heights reached by a rocket during a mission.

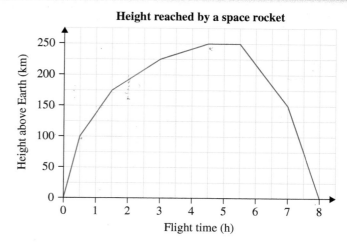

Height reached by a space rocket

a What is the maximum height reached by the rocket?

b How long is the rocket in flight?

c Which part of the journey is the fastest?

d Approximately how high is the rocket 2 hours after take-off?

5 Graphs are used to show other changes over time.

This graph shows the temperature reached during a chemical reaction.

Temperature during a reaction

a What is the highest temperature reached during the reaction?

b After how long is this highest temperature reached?

c What is the temperature at the start of the reaction?

d After $3\frac{1}{2}$ hours the temperature starts to fall.

How long does it take to get back to the starting temperature?

> Read from this graph in the same way. For example, after $\frac{1}{2}$ hour the temperature is 100°C.

Review

1 a A film lasts 2 hours and 53 minutes. Work out how many minutes this is.

  b If the film begins at 7.45pm, when does it end?

2 Write these times using the 24-hour clock.

  a 7.51am   **b** 6.12pm   **c** 12 noon   **d** 3.35pm

3 Write these times using the 12-hour clock.

  a 02 45   **b** 19 55   **c** 00 10   **d** 13 28

4 A plane takes off at 11.23am and lands at 5.12pm.

Work out how long the flight lasts.

5 A train journey lasts 3 hours and 43 minutes.

If the train departs at 14 55, when does it arrive at its destination?

6 The time in Sofia is $3\frac{1}{2}$ hours behind Yangon. Work out the time in Sofia when it is 21 04 in Yangon.

7 This is part of a timetable for a tourist bus around a city.

	Tour bus departures			
Railway station	11 24	12 24	13 15	14 23
Museum	11 36	12 36	13 27	14 35
Art gallery	11 45	12 45	13 36	14 44
Shopping mall	11 51	12 51	13 42	14 50
Tourist information	12 06	–	13 57	–
Grand hotel	12 19	–	1411	–
Market	12 27	–	14 20	–
Bus station	12 33	13 25	14 26	15 18

a Work out how many minutes it takes for the 11 24 bus from the railway station to reach the bus station.

b Work out which bus does the journey in the shortest time.

c Karl just misses the 12 51 bus from the shopping mall. How long will he wait for the next one?

d Anna leaves the art gallery at 12 05. Work out how long she will need to wait for a bus to the market.

8 Melvyn goes out for an afternoon drive in his car. The graph shows his trip.

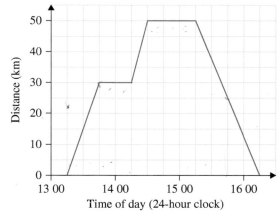

Melvyn's afternoon drive

a What time does he leave home?

b How far does he drive altogether?

c How long does he stop for altogether?

d How far from home is he at 3.45pm?

21 Sets ★

- Use set language and notation.

Animal sets

Set theory is so important that it is considered to be a foundation from which nearly all mathematics can be derived.

Mathematicians are not the only people that use sets. Scientists also use sets.

There are often many variations within a species of animal.

Some possible variations for frogs are:

- large or small
- warty skin or smooth skin
- webbed feet or not webbed feet
- brightly coloured skin or not brightly coloured skin
- poisonous or not poisonous
- with teeth or without teeth.

These variations are used when classifying frogs into different sets.

21.1 Describing sets

A set is a collection of items.

The items can be numbers, objects or symbols.

The items in the set are called the members or elements of the set.

The list of members is enclosed by { }.

Commas are used to separate the members.

For example {factors of 8} = {1, 2, 4, 8}

Some sets have no members.

The symbol $\varnothing$, or { } is used to represent an empty set.

For example {flying elephants} = $\varnothing$

Worked example

List the members of: **a** {prime numbers between 0 and 20} **b** {factors of 20}.

a {prime numbers between 0 and 20} = {2, 3, 5, 7, 11, 13, 17, 19}

b $1 \times 20 = 20$ $\qquad$ $2 \times 10 = 20$ $\qquad$ $4 \times 5 = 20$

{factors of 20} = {1, 2, 4, 5, 10, 20}

Exercise 21.1

1 Describe these sets in words.

  a {2, 4, 6, 8, ...}   **b** {1, 2, 3, 4, 5}

  c {1, 2, 3, 4, 6, 12}   **d** {North, East, South, West}

  e {2, 3, 5, 7, 11, 13, ...}   **f** {Monday, Tuesday, Wednesday, ...}

  g {7, 14, 21, 28, 35, ...}   **h** {1, 4, 9, 16, 25, ...}

  i {+, −, ×, ÷}   **j** {X, Y, Z}

2 List these sets.

  a {months of the year}   **b** {multiples of 3}

  c {prime numbers between 20 and 30}   **d** {square numbers between 50 and 150}

  e {first five letters of the alphabet}   **f** {odd numbers between 30 and 40}

  g {multiples of 6 between 40 and 70}   **h** {subjects you study at school}

  i {days of the week beginning with S}   **j** {names of quadrilaterals}

  k {factors of 15}   **l** {factors of 36}

3 Which of these sets are empty sets?

  a {days of the week that begin with P}   **b** {prime number that are even}

  c {multiples of 7 that are even}   **d** {square numbers between 70 and 80}

  e {years with 366 days}   **f** {countries south of the South Pole}

4 Describe these sets in words.

  a {A, B, C, D, E, H, I, K, M, O, T, U, V, W, X, Y}

  b {H, I, N, O, S, X, Z}

  c {101, 111, 121, 131, 141, 151, 161, 171, 181, 191}

  d {225, 256, 289, 324, 361}

  e {53, 59, 61, 67, 71, 73, 79, 83, 89, 97}

5 Which of these sets are empty sets? Explain your answers.

    a {quadrilaterals with two reflex angles}   **b** {triangles with two right angles}

    c {pentagons with five reflex angles}   **d** {quadrilaterals with four acute angles}

    e {hexagons with two reflex angles}   **f** {pentagons with three reflex angles}

21.2 Intersection and union of sets

The sets A = {2, 4, 6, 8, 10} and B = {4, 8, 12} can be shown on a diagram.

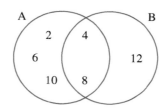

A ∩ B is the **intersection** of A and B and it contains the members that are in both A and B.

A ∩ B = {4, 8}

A ∩ B can be shaded on a diagram as:

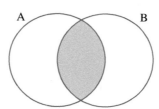

A ∪ B is the **union** of A and B and it contains the members that are in A or B or both.

A ∪ B = {2, 4, 6, 8, 10, 12}

A ∪ B can be shaded on a diagram as:

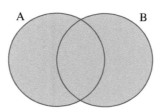

Worked example

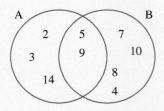

List the members of: **a** A   **b** B   **c** A ∩ B   **d** A ∪ B.

a A = {2, 3, 5, 9, 14}   select the numbers in the circle A

b B = {4, 5, 7, 8, 9, 10}   select the numbers in the circle B

c A ∩ B = {5, 9}   select the numbers that are in both A and B

d A ∪ B = {2, 3, 4, 5, 7, 8, 9, 10, 14}   select the numbers in A or B or both

Exercise 21.2

1 A group of students studies biology (B) or physics (P) or both.

The table shows this information.

Name	Anu	Ben	Coe	Dan	Eva	Fred	Gia
Biology (B)	✓	✗	✓	✓	✓	✓	✗
Physics (P)	✗	✓	✓	✗	✓	✗	✓

Copy and complete the diagram to show
this information.

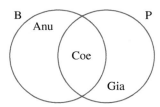

2 List the members of:

    a A   **b** B

    c A ∩ B   **d** A ∪ B.

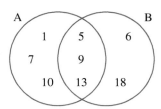

3 List the members of:

    a P   **b** Q

    c P ∩ Q   **d** P ∪ Q.

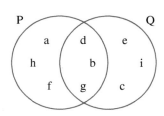

4 A = {multiples of 2 that are less than 20}

B = {multiples of 3 that are less than 20}

a List the members of: **i** A   **ii** B

b Show the sets A and B on a diagram.

c List the members of: **i** A ∩ B   **ii** A ∪ B

d Describe the set A ∩ B.

5 Find: **i** A ∩ B   **ii** A ∪ B   for each of these.

a A = {1, 3, 6, 7, 8}   B = {2, 3, 4, 5, 8}

b A = {b, c, d, f}   B = {a, b, e}

c A = {3, 4, 7, 10}   B = {2, 5, 6, 8, 9}

d A = {📞, 👓, ✉, ✏}   B = {📞, ✂, ✉, ✏}

e A = {▲, ▲, ■, ∕, ∕, ⬟, ⬢}   B = {▲, ■, ■, ∕, ⬟, ⬢}

6 On copies of this diagram shade the sets:

a C   **b** D

c C ∩ D   **d** C ∪ D

7 A group of students studies one or more of art (A), biology (B) and chemistry (C).

The table shows this information.

Name	Ana	Bev	Carl	Dave	Eve	Fred	Gary
Art (A)	✗	✗	✓	✓	✓	✓	✗
Biology (B)	✓	✓	✓	✗	✗	✓	✗
Chemistry (C)	✗	✓	✓	✗	✓	✗	✓

Copy and complete the diagram to show this information.

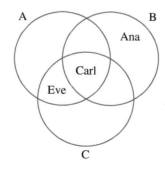

8 Look at the diagram opposite. List the members of:

a A ∩ B   **b** A ∩ C

c B ∩ C   **d** A ∩ B ∩ C

e A ∪ B   **f** A ∪ C

g B ∪ C   **h** A ∪ B ∪ C

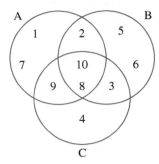

9 Look at the diagram opposite. List the members of:

a A ∩ B   **b** A ∩ C

c B ∩ C   **d** A ∩ B ∩ C

e A ∪ B   **f** A ∪ C

g B ∪ C   **h** A ∪ B ∪ C

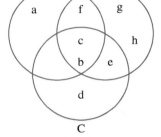

10 On copies of the diagram opposite shade:

a A   **b** C

c A ∩ B   **d** B ∩ C

e A ∩ B ∩ C   **f** A ∪ C

g B ∪ C   **h** A ∪ B ∪ C

i A ∩ (B ∪ C)   **j** A ∪ (B ∩ C)

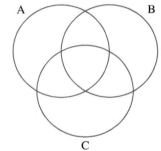

21.3 Elements and subsets

The number of elements in the set A is written as n(A).

If A = {2, 4, 6, 8, 10}, then n(A) = 5

If B = {p, q, r, s}, then n(B) = 4

The symbol ∈ means 'is a member of'.

The symbol ∉ means 'is not a member of'.

If A = {5, 10, 15, 20, 25, 30}, then 5 ∈ A and 4 ∉ A.

A is called a **proper subset** of B if all the members of A are contained in the larger set B.

The symbol ⊂ means 'is a proper subset of'.

If A = {◆, ●} and B = {■, ◆, ♠, ●, ✦}, then A ⊂ B.

This can be shown on a diagram as:

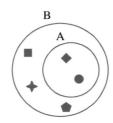

Worked example

Write down:

a n(B)   **b** n(C)   **c** n(A ∩ B)

d n(A ∪ C)   **e** n(B ∩ C)

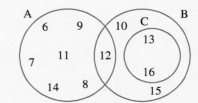

a B = {10, 12, 13, 15, 16}   n(B) = 5

b C = {13, 16}   n(C) = 2

c A ∩ B = {12}   n(A ∩ B) = 1

d A ∪ C = {6, 7, 8, 9, 11, 12, 13, 14, 16}   n(A ∪ C) = 9

e B ∩ C = {13, 16}   n(B ∩ C) = 2

Exercise 21.3

1 A = {3, 5, 7, 10}   B = {p, q, r}   C = {1, 3, 5, 9, 20, 31}   D = ∅

  a Find: **i** n(A)   **ii** n(B)   **iii** n(C)   **iv** n(D)

  b Which of these statements are true and which are false?

      i 8 ∈ A   **ii** r ∈ B   **iii** 9 ∈ C   **iv** 5 ∈ D

      v 3 ∈ B   **vi** 0 ∈ D   **vii** 5 ∈ A ∩ B   **viii** 9 ∈ A ∪ B

2

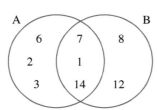

  Write down: **a** n(A)   **b** n(B)   **c** n(A ∩ B)   **d** n(A ∪ B)

3 Which of these statements are true and which are false?

    a $6 \in$ {odd numbers}   **b** green $\in$ {red, orange, yellow, green, blue}

    c $307 \in$ {prime numbers}   **d** $361 \in$ {square numbers}

4 Which of these statements are true and which are false?

    a {a, b, c} $\subset$ {a, b, c, d, e}   **b** {1, 4, 7, 10} $\subset$ {4, 10}

    c {3, 6} $\subset$ {2, 4, 6, 8}   **d** {April} $\subset$ {days of the week}

    e {factors of 10} $\subset$ {1, 2, 3, 4, 5}   **f** {1, 3, 5} $\subset$ {prime numbers}

    g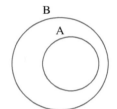

    h {◆, ◗, ●} $\subset$ {■, ◆, ◗, ♠, ●}

5 A = {3, 5, 7, 9}   B = {1, 2, 3, 4, 5, 6, 7, 8, 9}

Copy the diagram and put the elements in the correct places.

B
A
(diagram: circle A inside circle B)

6

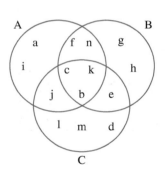

Write down:

    a n(A)   **b** n(B)   **c** n(C)   **d** n(A $\cap$ B)

    e n(A $\cap$ C)   **f** n(B $\cap$ C)   **g** n(A $\cap$ B $\cap$ C)   **h** n(A $\cup$ B)

    i n(A $\cup$ C)   **j** n(B $\cup$ C)   **k** n(A $\cup$ B $\cup$ C)

7 a 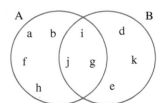   **b** 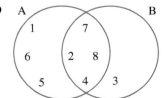   **c**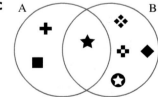

For each of the diagrams write down: **i** n(A $\cup$ B)   **ii** n(A)   **iii** n(B)   **iv** n(A $\cap$ B)

Write down a rule connecting n(A $\cup$ B), n(A), n(B) and n(A $\cap$ B).

Review

1 A = {factors of 40}

    a List the members of A.   **b** Find n(A).

2 Which of these sets are empty sets? Explain your answers.

    a {square numbers between 130 and 140}

    b {hexagons with three reflex angles}

    c {prime numbers between 290 and 300}

3 a List the members of: **i** P   **ii** Q

                       iii P ∩ Q   **iv** P ∪ Q

    b Find: **i** n(P)   **ii** n(Q)

                       iii n(P ∩ Q)   **iv** n(P ∪ Q)

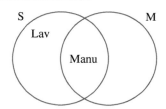

4 A = {1, 2, 3, 4, 7, 8, 9}   B = {4, 5, 6, 7, 10}

    a Find: **i** A ∩ B   **ii** A ∪ B

    b Show the sets A and B on a diagram.

5 A = {even numbers)   B = {prime numbers}

    List the members of A ∩ B.

6 The table shows the clubs that a group of students attend.

Name	Lav	Manu	Nila	Oma	Pari	Reva	Sai
Sports club (S)	✓	✓	✗	✗	✗	✓	✓
Music club (M)	✗	✓	✓	✓	✓	✗	✓

Copy and complete the diagram
to show this information.

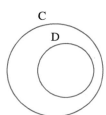

7 Which of these statements are true and which are false?

    a 91 ∈ {prime numbers}   **b** {factors of 12} ⊂ {factors of 36}

    c ✦ ∈ {✚, ★, ✿, ❖}   **d** {+, −, ×, ÷} ⊂ { ×, +, −}

8 C = {1, 2, 4, 5, 7, 8, 9} and D = {7, 8}

Copy the diagram and put the elements
in the correct places.

22 Matrices

Using matrices to organise data

Matrices are used for organising and dealing with large quantities of data.

Planning travel between different cities and countries can be very complicated, especially if there is more than one route.

A matrix is a useful and logical way of storing all the information.

Airlines, bus companies and train companies use matrices.

22.1 Route matrices

The diagram shows a network of roads connecting four towns A, B, C and D.

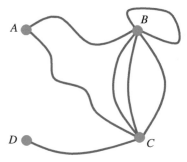

Information about the number of routes between each town (without passing through any other town) can be stored in a matrix.

$$
\begin{array}{c}
 & \text{To} \\
 & \begin{array}{cccc} A & B & C & D \end{array} \\
\text{From}\quad
\begin{array}{c} A \\ B \\ C \\ D \end{array}
&
\begin{pmatrix}
0 & 1 & 1 & 0 \\
1 & 2 & 3 & 0 \\
1 & 3 & 0 & 1 \\
0 & 0 & 1 & 0
\end{pmatrix}
\end{array}
$$

There are 3 routes from B to C.

There are 2 routes from B to B. (You can go clockwise or anticlockwise around the ring road.)

If there is an arrow on a road it means that you can only travel along the road in the direction of the arrow. (It is called a one-way road.)

If there is no arrow on a road it means that you can travel in both directions along the road.

Worked example

Write down the route matrix for this network of roads.

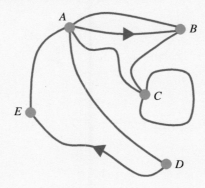

The route matrix is:

$$
\begin{array}{c c}
 & \text{To} \\
 & \begin{array}{ccccc} A & B & C & D & E \end{array} \\
\text{From} \quad
\begin{array}{c} A \\ B \\ C \\ D \\ E \end{array}
&
\left(
\begin{array}{ccccc}
0 & 2 & 1 & 1 & 1 \\
1 & 0 & 1 & 0 & 0 \\
1 & 1 & 2 & 0 & 0 \\
1 & 0 & 0 & 0 & 1 \\
1 & 0 & 0 & 0 & 0
\end{array}
\right)
\end{array}
$$

Note: There are two one-way roads.

There is only one route from B to A.

There is no route from E to D.

Exercise 22.1

1 Write down a route matrix for each of these networks.

a

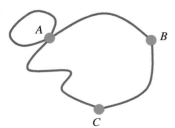

b

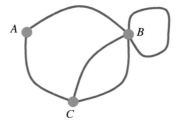

c

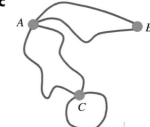

d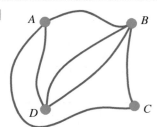

e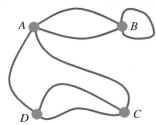

f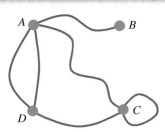

g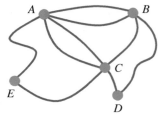

h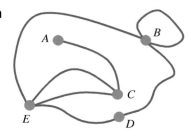

i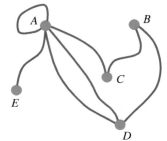

2 Draw a network for each of these route matrices.

a From $\begin{array}{c} \\ A \\ B \\ C \end{array} \begin{array}{c} \overset{\text{To}}{} \\ \begin{array}{ccc} A & B & C \end{array} \\ \begin{pmatrix} 0 & 1 & 2 \\ 1 & 2 & 0 \\ 2 & 0 & 2 \end{pmatrix} \end{array}$

b From $\begin{array}{c} \\ A \\ B \\ C \end{array} \begin{array}{c} \overset{\text{To}}{} \\ \begin{array}{ccc} A & B & C \end{array} \\ \begin{pmatrix} 0 & 0 & 1 \\ 0 & 2 & 1 \\ 1 & 1 & 0 \end{pmatrix} \end{array}$

c From $\begin{array}{c} \\ A \\ B \\ C \end{array} \begin{array}{c} \overset{\text{To}}{} \\ \begin{array}{ccc} A & B & C \end{array} \\ \begin{pmatrix} 2 & 1 & 0 \\ 1 & 0 & 3 \\ 0 & 3 & 0 \end{pmatrix} \end{array}$

d From $\begin{array}{c} \\ A \\ B \\ C \\ D \end{array} \begin{array}{c} \overset{\text{To}}{} \\ \begin{array}{cccc} A & B & C & D \end{array} \\ \begin{pmatrix} 0 & 1 & 1 & 0 \\ 1 & 0 & 2 & 0 \\ 1 & 2 & 2 & 1 \\ 0 & 0 & 1 & 0 \end{pmatrix} \end{array}$

e From $\begin{array}{c} \\ A \\ B \\ C \\ D \end{array} \begin{array}{c} \overset{\text{To}}{} \\ \begin{array}{cccc} A & B & C & D \end{array} \\ \begin{pmatrix} 2 & 3 & 1 & 0 \\ 3 & 0 & 2 & 0 \\ 1 & 2 & 0 & 1 \\ 0 & 0 & 1 & 2 \end{pmatrix} \end{array}$

f From $\begin{array}{c} \\ A \\ B \\ C \\ D \\ E \end{array} \begin{array}{c} \overset{\text{To}}{} \\ \begin{array}{ccccc} A & B & C & D & E \end{array} \\ \begin{pmatrix} 0 & 0 & 1 & 2 & 1 \\ 0 & 0 & 0 & 1 & 0 \\ 1 & 0 & 0 & 0 & 1 \\ 2 & 1 & 0 & 0 & 1 \\ 1 & 0 & 1 & 1 & 2 \end{pmatrix} \end{array}$

3 Write down a route matrix for each of these networks.

a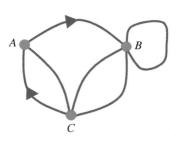

b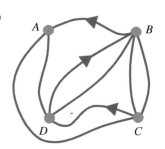

c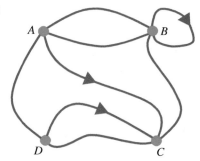

22.2 Adding and subtracting matrices

These tables show the number of gold (G), silver (S) and bronze (B) medals won by China and the USA in the 2008 and 2012 Summer Olympic Games.

2008	G	S	B
China	51	21	28
USA	36	38	36

2012	G	S	B
China	38	27	23
USA	46	29	29

The tables can be combined to give the total number of gold, silver and bronze medals for China and the USA.

Total	G	S	B
China	89	48	51
USA	82	67	65

This can be written in matrix form as $\begin{pmatrix} 51 & 21 & 28 \\ 36 & 38 & 36 \end{pmatrix} + \begin{pmatrix} 38 & 27 & 23 \\ 46 & 29 & 29 \end{pmatrix} = \begin{pmatrix} 89 & 48 & 51 \\ 82 & 67 & 65 \end{pmatrix}$

You can add two matrices if they are the same size.

The size of a matrix is called the **order** of the matrix.

The matrix $\begin{pmatrix} 51 & 21 & 28 \\ 36 & 38 & 36 \end{pmatrix}$ has order 2 × 3. (2 rows and 3 columns.) — The number of rows is always written first.

Worked example 1

$\begin{pmatrix} 3 & 1 & 5 & 0 & 2 \\ 4 & 2 & 1 & 1 & 0 \\ 2 & 0 & 2 & 5 & 1 \end{pmatrix}$

Write down the order of the matrix.

$\begin{pmatrix} 3 & 1 & 5 & 0 & 2 \\ 4 & 2 & 1 & 1 & 0 \\ 2 & 0 & 2 & 5 & 1 \end{pmatrix}$

The matrix has 3 rows and 5 columns. The order is 3 × 5

Worked example 2

$A = \begin{pmatrix} 6 & -1 & 3 \\ 0 & 7 & 2 \end{pmatrix}$ $B = \begin{pmatrix} 4 & 1 & 3 \\ 2 & 4 & -1 \end{pmatrix}$

Find: **a** A + B   **b** A − B   **c** A + A

a $A + B = \begin{pmatrix} 6+4 & -1+1 & 3+3 \\ 0+2 & 7+4 & 2+-1 \end{pmatrix} = \begin{pmatrix} 10 & 0 & 6 \\ 2 & 11 & 1 \end{pmatrix}$

b $A - B = \begin{pmatrix} 6-4 & -1-1 & 3-3 \\ 0-2 & 7-4 & 2--1 \end{pmatrix} = \begin{pmatrix} 2 & -2 & 0 \\ -2 & 3 & 3 \end{pmatrix}$

c $A + A = \begin{pmatrix} 6+6 & -1+-1 & 3+3 \\ 0+0 & 7+7 & 2+2 \end{pmatrix} = \begin{pmatrix} 12 & -2 & 6 \\ 0 & 14 & 4 \end{pmatrix}$

Exercise 22.2

1 Write down the order of these matrices.

a $(2 \quad 3 \quad 1)$ **b** $(8 \quad 2)$ **c** $(4 \quad 2 \quad 1 \quad 8)$ **d** (3)

e $\begin{pmatrix} 3 & 0 \\ 5 & 1 \end{pmatrix}$ **f** $\begin{pmatrix} 1 & 5 \\ 2 & 9 \\ 8 & 4 \end{pmatrix}$ **g** $\begin{pmatrix} 1 & 0 & 0 & 3 \\ 5 & 1 & 4 & 2 \end{pmatrix}$ **h** $\begin{pmatrix} 5 \\ 4 \\ 5 \end{pmatrix}$

i $\begin{pmatrix} 7 & 0 & 1 \\ 5 & 1 & 2 \\ 0 & 6 & 3 \\ 2 & 8 & 1 \end{pmatrix}$ **j** $\begin{pmatrix} 2 & 2 & 1 & 1 & 0 \\ 0 & 3 & 1 & 3 & 0 \\ 1 & 0 & 1 & 1 & 2 \end{pmatrix}$ **k** $\begin{pmatrix} 3 \\ 8 \\ 0 \\ 4 \end{pmatrix}$ **l** $\begin{pmatrix} 4 & 1 & 5 \\ 0 & 8 & 2 \end{pmatrix}$

2 $A = \begin{pmatrix} 5 & 2 & 1 & 8 \\ 3 & 9 & 7 & 6 \\ 4 & 7 & 9 & 2 \end{pmatrix}$ $B = \begin{pmatrix} 2 & 1 & 1 & 4 \\ 2 & 5 & 3 & 5 \\ 1 & 3 & 6 & 2 \end{pmatrix}$

Find:

a $A + B$ **b** $A - B$ **c** $A + A$ **d** $B + B + B$

3 $A = \begin{pmatrix} 3 & 7 \\ 0 & 5 \end{pmatrix}$ $B = \begin{pmatrix} 2 & 4 \\ 6 & 8 \end{pmatrix}$ $C = \begin{pmatrix} 5 & 4 \\ -1 & 6 \end{pmatrix}$

Find:

a $A + B$ **b** $A + C$ **c** $B + C$

d $A - B$ **e** $A - C$ **f** $B - C$

g $A + A$ **h** $B + B + B$ **i** $C + C + C + C$

4 $A = (8 \quad 6 \quad 7)$ $B = (2 \quad 5 \quad 1)$ $C = (0 \quad -2 \quad 1)$

Find:

a $A + B$ **b** $A + C$ **c** $B + C$

d $A - B$ **e** $A - C$ **f** $B - C$

g $A + A + A$ **h** $B + B + B + B$ **i** $C + C + C$

5 $\begin{pmatrix} 5 & x \\ 1 & 0 \end{pmatrix} + \begin{pmatrix} 3 & 7 \\ y & 8 \end{pmatrix} = \begin{pmatrix} 8 & 10 \\ 5 & 8 \end{pmatrix}$

Find the values of x and y.

6 $\begin{pmatrix} 4 & x & 8 & 9 \\ 7 & 5 & 1 & 6 \\ 2 & 4 & z & 0 \end{pmatrix} + \begin{pmatrix} 3 & 7 & 3 & 6 \\ 5 & 0 & 1 & y \\ -1 & 8 & 4 & 3 \end{pmatrix} = \begin{pmatrix} 7 & 10 & 11 & 15 \\ 12 & 5 & 2 & 4 \\ 1 & 12 & 7 & 3 \end{pmatrix}$

Find the values of x, y and z.

22.3 Multiplying a matrix by a number

If $A = \begin{pmatrix} 5 & 2 \\ 3 & 1 \end{pmatrix}$ then:

$2A = A + A = \begin{pmatrix} 5 & 2 \\ 3 & 1 \end{pmatrix} + \begin{pmatrix} 5 & 2 \\ 3 & 1 \end{pmatrix} = \begin{pmatrix} 10 & 4 \\ 6 & 2 \end{pmatrix}$

2A is the same as multiplying each number in the matrix A by the number 2.

Worked example

$A = \begin{pmatrix} 6 & 8 & 4 \\ 1 & 5 & 7 \end{pmatrix}$ Find 3A.

$3A = \begin{pmatrix} 3 \times 6 & 3 \times 8 & 3 \times 4 \\ 3 \times 1 & 3 \times 5 & 3 \times 7 \end{pmatrix} = \begin{pmatrix} 18 & 24 & 12 \\ 3 & 15 & 21 \end{pmatrix}$

Exercise 22.3

1 $A = (3 \quad 1 \quad 5)$ Find 3A.

2 $B = \begin{pmatrix} 1 & 5 \\ 2 & 4 \end{pmatrix}$ Find 4B.

3 $C = \begin{pmatrix} 2 & 0 & 1 & 4 \\ 3 & 5 & 6 & 1 \\ 1 & 3 & 5 & 2 \end{pmatrix}$ Find 7C.

4 $A = \begin{pmatrix} 2 & 4 \\ 0 & 1 \end{pmatrix}$ $B = \begin{pmatrix} 1 & 0 \\ 2 & 3 \end{pmatrix}$ $C = \begin{pmatrix} 3 & 1 \\ 1 & 5 \end{pmatrix}$

Find:

   a 2A + B   **b** B + 3C   **c** 3A + 2C   **d** 4C − 3B

Review

1 Write down a route matrix for each of these networks.

a

b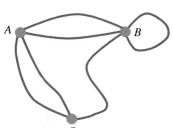

c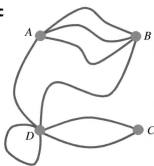

2 Draw a network for each of these route matrices.

a From $\begin{matrix} & \overset{\text{To}}{\begin{matrix} A & B \end{matrix}} \\ \begin{matrix} A \\ B \end{matrix} & \begin{pmatrix} 2 & 2 \\ 2 & 0 \end{pmatrix} \end{matrix}$

b From $\begin{matrix} & \overset{\text{To}}{\begin{matrix} A & B & C \end{matrix}} \\ \begin{matrix} A \\ B \\ C \end{matrix} & \begin{pmatrix} 2 & 1 & 2 \\ 1 & 2 & 0 \\ 2 & 0 & 0 \end{pmatrix} \end{matrix}$

c From $\begin{matrix} & \overset{\text{To}}{\begin{matrix} A & B & C & D \end{matrix}} \\ \begin{matrix} A \\ B \\ C \\ D \end{matrix} & \begin{pmatrix} 0 & 0 & 1 & 2 \\ 0 & 2 & 1 & 3 \\ 1 & 1 & 0 & 1 \\ 2 & 3 & 1 & 2 \end{pmatrix} \end{matrix}$

3 Write down a route matrix for this network.

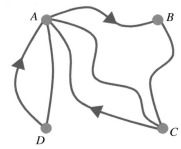

4 Write down the order of these matrices.

a $\begin{pmatrix} 3 \\ 2 \end{pmatrix}$
b $(1 \quad 5 \quad 7)$
c $\begin{pmatrix} 5 & 1 \\ 0 & 2 \end{pmatrix}$
d $\begin{pmatrix} 2 & 5 & 4 \\ 1 & 0 & 7 \end{pmatrix}$

e $\begin{pmatrix} 5 \\ 2 \\ 8 \end{pmatrix}$
f $\begin{pmatrix} 1 & 4 & 5 & 0 \\ 2 & 2 & 7 & 3 \\ 1 & 8 & 0 & 5 \end{pmatrix}$
g $(2 \quad 0 \quad 9 \quad 5)$
h $\begin{pmatrix} 1 & 0 \\ 0 & 6 \\ 2 & 5 \\ 3 & 3 \end{pmatrix}$

5 $A = \begin{pmatrix} 2 & 0 \\ 6 & 4 \end{pmatrix}$
$B = \begin{pmatrix} 5 & 1 \\ -2 & 3 \end{pmatrix}$
$C = \begin{pmatrix} -3 & 0 \\ 2 & 7 \end{pmatrix}$

Find:

a $A + B$
b $B + C$
c $A - C$
d $A + B + C$

e $2A$
f $3B$
g $4C$
h $2A - B$

6 $\begin{pmatrix} x & 8 \\ -2 & 3 \end{pmatrix} + \begin{pmatrix} -5 & -2 \\ 4 & y \end{pmatrix} = \begin{pmatrix} -1 & 6 \\ 2 & -4 \end{pmatrix}$

Find the values of x and y.

Glossary

12-hour clock A system of recording time that repeats every 12 hours (twice each day).

24-hour clock A system of recording time that repeats every 24 hours (once each day).

A

acute angle An angle between 0° and 90°.

area A measure of the size of the surface of a shape.

axes The plural of axis.

B

brackets Brackets are used to group terms together.

C

capacity The amount of space inside a body. The amount it can hold.

centimetre (cm) A metric unit that is equal to $\frac{1}{100}$ of a metre.

centre of rotation The point about which a rotation takes place.

common factor The common factors of two numbers are those integers which are factors of both of them.

common multiple A common multiple of two numbers is a number that is a multiple of both of them.

congruent Identical in a shape and size.

construction line A faint line used in construction. Although not part of the final answer, it should be left visible to show how the construction was done.

coordinates A pair of numbers used to show a position on a graph.

cube A solid (polyhedron) with six square faces.

cubic centimetre A unit of volume equal to a cube 1 cm by 1 cm by 1 cm.

cubic metre A unit of volume equal to a cube 1 m by 1 m by 1 m.

cubic millimetre A unit of volume equal to a cube 1 mm by 1 mm by 1 mm.

cuboid A solid (polyhedron) with six rectangular faces.

D

data collection sheet A form used to record data as it is collected.

day A unit of time equal to 24 hours.

denominator The name given to the number on the 'bottom' of a fraction. It tells you how many parts the whole has been divided into.

derive Write your own formula to connect some variables.

diagonal A line within a polygon joining two vertices.

direct proportion Two quantities are in direct proportion if they are in a constant ratio.

directed number A number with a + or − sign showing whether it is positive or negative.

distance–time graph Another name for a travel graph.

distribution A group of values.

E

edge A line formed where two faces of a solid object meet.

element (member) An item that belongs to a set.

empty set A set that contains no elements.

equally likely outcomes Outcomes that have the same probability.

equation A mathematical statement that says that two expressions are equal.

equilateral triangle A triangle with three equal sides and three equal angles.

equivalent fraction A fraction that has the same value as another fraction.

event A set of outcomes in probability.

expanding brackets Multiplying each term inside the brackets by the term outside the brackets.

experiment A series of trials.

experimental probability The probability of a chosen outcome. It is found by carrying out a number of trials.
Experimental probability = $\frac{\text{number of successful outcomes}}{\text{total number of trials}}$

expression   A collection of terms that does not include an equal sign.

F

face   A flat or curved surface of a solid object.

factor   The factors of a number are those integers that divide exactly into the number.

formula   A formula is a rule that connects two or more variables.

frequency diagram
Any type of diagram that shows frequencies, for example a bar chart.

frequency table   A table showing the number of times each value occurs.

function   A rule connecting two sets of numbers.

function machine   A diagram that shows the steps in a function.

G

gram (g)   A small unit for mass in the metric system.

H

hexagon   A polygon with six sides.

highest common factor
The highest common factor is the largest number that is a common factor of two or more given numbers.

hour   A unit of time equal to 60 minutes.

I

image   The final shape after a transformation.

improper fraction   Any fraction in which the numerator is greater than the denominator.

input   The number that goes into a function machine.

integer   An integer is a whole number.

internal angle   The angle between two sides of a polygon which meet at a vertex.

intersection   The intersection of sets A and B is the set that contains the elements that are in both A and B.

inverse   One operation that reverses the effect of another.

isosceles triangle   A triangle with two equal sides and two equal angles.

K

kilogram (kg)   A metric unit equal to 1000 grams.

kilometre (km)   A metric unit equal to 1000 metres.

kite   A quadrilateral with two pairs of equal sides but no parallel sides.

L

length   The measurement between two points.

like term   A term containing the same variables as another term.

line of symmetry   A line in which a shape reflects on to itself.

line symmetry   A shape has line symmetry if it is unchanged when reflected in a line.

litre (l)   A unit for capacity in the metric system.

lowest common multiple
The lowest common multiple of two or more numbers is the smallest multiple of all of them.

lowest terms (also called simplest form)   A fraction that cannot be simplified any more.

M

mapping diagram   A diagram that shows which numbers the inputs map to.

mass   A measure of the amount of matter in a body. Closely connected to, but not quite the same as the weight of a body.

matrix   A matrix is a rectangular arrangement of numbers.

mean   The mean is found by adding all the values together and dividing the total by the number of values. It is used as an average value.

median   The middle value when all the values are listed in order of size. It is used as an average value.

member (element)   An item that belongs to a set.

metre (m)   A unit for length in the metric system.

millilitre (ml)   A metric unit equal to $\frac{1}{1000}$ of a litre.

millimetre (mm)   A metric unit equal to $\frac{1}{1000}$ of a metre.

minute   A unit of time equal to 60 seconds.

mirror line   The line acting as the mirror for a reflection.

mixed number   A number that has both a whole number part and a fraction part.

modal class   The class (group) with the highest frequency.

mode   The value that occurs most frequently. It is used as an average value.

multiple   A multiple of k is a number that k divides into exactly.

N

negative   A negative number is less than 0.

net   A two-dimensional shape that can be folded to create a three-dimensional object.

number line   A line showing positive integers to the right of zero and negative integers to the left.

numerator   The name given to the number on the 'top' of a fraction. It tells you how many of the equal parts you have or are interested in.

O

object   The original shape before a transformation.

obtuse angle   An angle between 90° and 180°.

octagon   A polygon with eight sides.

operation   An operation is a single step in a calculation, such as addition or multiplication.

order   The order of a matrix is the number of rows and columns that a matrix has. For example, 2 × 3 means 2 rows and 3 columns.

origin   The point where the axes of a graph cross.

outcomes   Results.

output   The number that comes out of a function machine.

P

parallel   Two or more lines that never meet or intersect.

parallelogram   A quadrilateral with two pairs of parallel sides.

pentagon   A polygon with five sides.

perimeter   The distance around the edge of a shape.

perpendicular   At right angles.

pie chart   A method of showing information on a circle. The circle is divided into sectors. The angle of each sector is in proportion to the fraction of the total data it shows.

polygon   A two-dimensional shape with straight sides.

polyhedron   A three-dimensional solid with polygon faces.

positive   A positive number is greater than 0.

prime number   A number with exactly two factors, 1 and itself.

prism   A polyhedron with a constant cross-section.

proper subset   A is a proper subset of B if all the elements of A are contained in the larger set B.

proportion   A comparison of two or more ratios.

pyramid   A polyhedron with a polygon base and triangular sides leading up to a point.

Q

quadrilateral   A shape with four straight sides.

questionnaire   A collection of questions used to collect data.

R

range   The range is the difference between the largest and smallest values. It is used as a measure of spread.

ratio   The relative size of two or more quantities.

rectangle   A parallelogram with four right angles.

reflection   A transformation that makes a mirror image of the object shape.

reflex angle   An angle between 180° and 360°.

regular polygon   A polygon with equal sides and angles.

relative frequency   The proportion of outcomes of an experiment that give a chosen result.
Relative probability = $\frac{\text{number of successful outcomes}}{\text{total number of trials}}$

rhombus   A parallelogram with equal sides.

right angle   An angle equal to 90°.

right-angled triangle   A triangle with one angle equal to 90°.

rotation   A transformation involving turning about a point.

rotation symmetry   A shape has rotation symmetry if it is unchanged when rotated.

S

scalene triangle   A triangle with three unequal sides and three unequal angles.

second   A short unit of time.

sequence   A list of numbers or diagrams that are connected by a rule.

set   A collection of items.

simplest form (also called lowest terms)   A fraction that cannot be simplified any more.

solve   To work out the correct answer.

square   A quadrilateral with four equal sides and four right angles.

square centimetre   A unit of area equal to a square 1 centimetre by 1 centimetre.

square metre   A unit of area equal to a square 1 metre by 1 metre.

square millimetre   A unit of area equal to a square 1 millimetre by 1 millimetre.

square number   A square number is a number formed by multiplying any integer by itself.

square root   The square root of a number is the number which, when multiplied by itself, is equal to the original number.

substitution   Replacing the letters in an expression or a formula by the given numbers.

surface area   The total area of the faces of a three-dimensional shape.

symmetry   A shape has symmetry if it is unchanged when reflected or rotated.

T

term   A number in a sequence.

term-to-term rule   A rule that connects a term with the previous term in a sequence.

theoretical probability   A probability based on equally likely outcomes.

timetable   A table showing the departure and arrival times of trains, buses or planes. Times are usually shown using the 24-hour clock.

tonne (t)   A metric unit equal to 1000 kilograms.

transformation   A movement of one shape to another.

translation   A transformation involving movement with no rotation or reflection.

transversal   A line that crosses two or more parallel lines.

trapezium   A quadrilateral with one pair of parallel sides.

travel graph   A graph showing distance travelled over a period of time.

trial   A single experiment, such as the roll of a dice.

U

union   The union of sets A and B is the set that contains the elements that are in A or B or both.

V

variable   A symbol, usually a letter such as x, that represents a quantity and that can take different values.

vertex   The point where two or more lines meet.

vertically opposite   Two angles created when two straight lines cross.

vertices   The plural of vertex.

volume   A measure of the space occupied by a three-dimensional solid.

X

x-axis   The number line on a graph used for showing horizontal position.

Y

y-axis   The number line on a graph used for showing vertical position.

Index

Key terms that appear in the glossary are in **bold**.

12-hour clock **266–7**
24-hour clock **267**

A
abundant numbers 19
acute angles **33**, 34, 122
adding 13, 16
  decimals 78, 79
  directed numbers 13
  large numbers 15
  negative numbers 13–14
algebra 20, 21
algebraic wall 25, 26
am (time) 267
angles 32–8
  acute **33**, 34, 122
  base 40, 41
  estimating 33
  external 39, 44
  interior (allied) 125
  internal (polygons) **137**
  labelling 36
  obtuse **33**, 34, 42, 122
  opposite 36–7, 122, 124
  at a point 35, 42, 150, 151
  reflex **33**, 127
  right **32**, 35, 123
  of rotation 187
  on a straight line 35, 36, 37, 39, 122
  of a triangle 39–40, 117
  vertically opposite **36**–7, 122
Angle-Side-Angle (ASA) 132–3
animal sets 277
area **157–9**
  of rectangles 158–9
  of compound shapes 163, 164–5
average 84, 87
  to compare values 93–6
  see also mean, median, mode
axes of graphs **193**

B
bar charts 145
bar-line graphs 145–6
base angles 40, 41
bicycle ride 272–3
BIDMAS 16, 178
brackets 16, **27**
  equations with 116
  expanding **27**, 116
buildings, tall 73
bus times 270

C
calculator, use of 12
capacity **105**

cement 255, 262
'centi' 100
centimetre (cm) **100**
centre of rotation **187**, 195
certain 208, 209, 211, 212
cheese 237, 239, 258
clock systems **266–7**
clocks 267
closed questions 246, 247
coins, fair 211, 212
collecting like terms 24, 27
common factors **6**, 7
  highest 6, 256, 257
common multiples **3**
  lowest 3
comparing
  distributions 93–6
  fractions 62
  average values 93–6
compound shapes 163–5
congruent **182**, 187, 191
construction lines **129**, 130, 133, 188
converting metric units 100, 101, 161
coordinate grid 193–6
coordinates **193–5**
cubes (solids) 139, 140, **167**–8
  volume 169
cubic centimetre **168**
cubic metre **168**
cubic millimetre **168**
cuboids 139, 140, **167**
  nets of 171
  surface area 170–1, 179
  volume 168–9

D
data 84–98, 243–5, 286
data collection sheet **244**–5
data, grouped and ungrouped 146, 249, 250
days **266**
decimal point 68–9
decimals 68–83
  adding 78, 79
  dividing 78
  to fractions 81
  multiplying 77
  subtracting 78
deficient numbers 19
denominator 54, **55**, 58, 63, 80, **232**
derive (formulae) **175–6**
designs, mathematical 121
diagonals **43**
dice, fair 212
digital scales 107

direct proportion **262**
directed numbers **12–14**
discrete data, grouped 146
distance–time formula 175
distance–time graph **273**
distribution **93–6**
dividing 16
  by 10, 100, 1000 75, 76
  decimals 78
  in given ratios 258–60
divisibility tests 3–4, 5, 7
drawing
  reflex angles 127
  hexagons 137–8
  lengths 125
  parallel lines 130
  perpendicular lines 129
  regular polygons 137–8
  rectangles 136
  squares 136
  triangles 131–3

E
edges **139**, 140
elements of sets **277–8**, 282, 283
empty sets **277**
equally likely outcomes 211, **212**
equations 111–120, 175
  with brackets 116
  of a line 225
  with one operation 111–13
  with two operations 114
  to solve problems 117
  solving **111**
equilateral triangles **40**, 49
equivalent fractions **56**, 58–9, 63, 233, 234
Eratosthenes, sieve of 8–9
Euclid 1
even chance 211, 212
even numbers 3
events 208–9
expanding brackets **27**, 116
experimental probability **216**–17
experiments 216–17
expressions 20–31, **17**
  simplifying 2
  substitution into
external angles 39,

F
faces **139**
factors 4
  common
  highest common 6, 25
  proper

Fibonacci sequence 202
first term (sequences) 201
formulae **175–81**
  area 159
  surface area 179
  deriving 175–6
  perimeter 175
  probability 212
  substitution into 178, 179
fractions 54–67
  adding 63
  of an amount 65, 232–3
  comparing 62
  to decimals 80–1, 237
  equivalent **56**, 58–9, 63, **233**, **234**
  improper **61–2**
  simplest form **59**, **233**
  subtracting 63
frequency diagrams **145**, 146–7
frequency tables 89–91, **145**, 146, 249–50
  mean from 91
frogs 277
function machines **221–3**
functions **221–3**

G

gram (g) **103**
graphs 143, 193–6, 221, 272–3
  bar-line 145–6
  straight-line 225, 227–8
  travel **273**
grouped discrete data 146

H

hexagonal prisms 140
hexagons 31, **49**, 137–8, 203
highest common factor **6**, 256, 257
horizontal lines 225
hours (h) **266**

I

image 182, 183, 184, 187, 191
impossible 208, 209, 211
improper fractions **61–2**
ices (powers) 16
t 221, 222, 223
ers **6**, 71
r (allied) angles 125
l angles 137
ion (sets) 279–80
n scales) 107
1, 2, 12
ations 112–13
eziums 49
gles 39–40, 41, 42, 49
140
6, 8
6, 7
6, 257
19
267–8, 270

K

'kilo' 100
kilogram (kg) **103**
kilometre (km) **100**
kites 43, 47, 49, 124

L

large numbers 15
length 99–100, 125
letters representing numbers 20
like terms **24**
  collecting 24, 27, 117
likely 208, 209
line symmetry **46–7**, 49
lines
  parallel 43, **121–2**, 130
  perpendicular **123**, 129
  of symmetry 46–7, **182**, 183, 184
litre (l) **105**
logos 138
lowest common multiple **3**
lowest terms 59, 233

M

mapping diagrams **222–3**
mass **102–4**
mathematical designs 121
mathematicians 1, 8, 19, 111, 208
matrices **286–92**
  adding 289–90
  multiplying by a number 291
  order of **289**
  subtracting 289–90
mean 87–8, 94, 95
  from frequency tables 91
measure of spread **88**, 95–6
measuring 99, 102, 105, 125
  angles 126–7
median 85–6, 94, 95
members (of sets) 277–8, 282, 283
metre (m) **99**
metric system 99–100
  converting units 100, 101, 161
'milli' 100
millilitre (ml) **105**
millimetre (mm) **100**
minutes (min) **266**
mirror lines 46–7, **182**, 183
  diagonal 184
mixed numbers **61–2**
modal class **146–7**
mode (modal number) **84–5**
multiples 1, 2–3
  common **3**
multiplying 16
  by 10, 100, 1000 74, 75
  decimals 77
  matrices by numbers 291

N

negative numbers **12–14**
nets **170–1**
Nicomachus 19
number lines **12–14**
number sequences 200–1
numbers
  abundant 19
  deficient 19
  large 15
  mixed **61–2**
  perfect 19
  square **11**
  whole 6, 71
numerator 54, **55**, 58, 63, 80, **232**

O

object **182**, 183, 184, 187, 191
obtuse angles 33, 34, 42, 122
octagons **49**, 121, 138, 142
octahedrons 140
'of' 65, 238
Olympic medals 289
open questions 246, 247
operations 2, 16, 222
  equations with one 111–13
  equations with two 114
  inverse 112–13
  order of 16
opposite angles 36–7, 122, 124
order of a matrix **289**
origin **193**, 194
outcomes 211, **212**, 216
output **221**, 222, 223

P

parallel lines 43, **121–2** , 130
parallelograms **43**, 49
part of an amount
  fractions 232–3
  percentages 234–5, 238–9
Pascal, Blaise 208
patterns of tiles 121, 142, 255
pentagonal prism 139
pentagons 23, 26, **49**, 138
per cent (%) 232
percentages 234–5, 238–9
perfect numbers 19
perimeter **160**
  of compound shapes 164, 165
  of rectangles 160, 175
perpendicular lines 123, 129
personal information 247
pictograms 143–4
pie charts **150–1**
pizzas 54
place value 68–9, 74–6
  tables 68–9, 74–6, 81
pm (time) 267

polygons   43, 121, **137–8**
polyhedrons   **139**
positive numbers   **12**, 13, 14
powers (indices)   16
prime number grid   9
prime numbers   1, **8–9**
prisms   **139**, 140
probability   208–20
  calculating   212–13
  estimating   216–17
  experimental   **216–17**
  formulae   212
  scale   211–12
  theoretical   **212**
  words   208–9
proper factors   19
proper subset   **283**
proportion   **261–2**
protractor   126, 127, 131, 133
pyramids   **139**, 140

Q

quadrilaterals   26, **43–5**
  angles in   43, 44
questionnaire design   **246–7**
questions   246

R

random   212, 213
range   88, **95–6**
ratio   **255**–7, 262
  dividing in given   258–60
  simplest form   255, 256, 257, 261
recipes   256, 258, 263, 265
Recorde, Robert   111
rectangles   26, **43**, 47, 49
  area   158–9
  drawing   136
  perimeter   160, 175
reflection   **182–4**
reflex angles   **33**, 127
regular polygons   **137–8**
relative frequency   **216**, 217
rhombuses   **43**, 49, 124
right angles   **32**, 35, 123
right-angled isosceles
  triangles   41
right-angled triangles   **40**, 41
road networks   286–7
rotation   **187**–8, 195
rotation symmetry   **46**, 48, 49
rounding   71–3
route matrices   286–7
ruler, using   125

S

scalene triangles   **40**, 49
scales   107, 111–12
  reading   106–7
seconds (s)   **266**

sequences   **200–7**
  Fibonacci   202
  of numbers   200–1
  from shapes   203–4
set squares   129, 130, 136
sets   **277–85**
  animal   277
  empty   277
shapes, compound   163–5
shapes, sequences from   203–4
Side–Angle–Side (SAS)   131
sieve of Eratosthenes   8–9
simplest form   **59**, 233
simplifying expressions   24
simplifying fractions   59
solids   139–40
solving equations   **111**, 117
speed–time formula   175
spread, measure of   88, 95–6
square centimetre   **157**, 161
square metre   **158**
square millimetre   **158**, 161
square numbers   **11**
square roots   **12**
squares (figures)   11, **43**, 49, 115,
                                   121, 136
statistics   243
straight lines   225, 227–8
  graphs of   225, 227–8
subset, proper   **283**
substitution
  into expressions   **29**
  into formulae   178, 179
subtracting   13, 14, 16
  decimals   78
  directed numbers   13, 14
  large numbers   15
  negative numbers   13, 14
surface area   **170**–1, 179
surveys   153, 246, 247
symbols   111, 282, 283
symmetry   **46–8**
  line   **46**–7, 49
  rotation   **46**, 48, 49

T

tally charts and marks   244
temperature   13, 14, 109
terms   **24**
  collecting like   24, 27, 117
  like   **24**
  lowest   **59**, 233
  of a sequence   **200**, 201
term-to-term rule   **200**, 201, 202
theodolite   32
theoretical probability   **212**
thermometers   13
tile patterns   121, 142, 255
time measurement   68, 70, 266
time, units of   266

time–distance formula   175
times of journeys   267–8, 270
timetables   **270**
tonne (t)   **103**
train times   267–8
transformation   **182–4**, **187–8**,
                              191–2, 195
translation   **191–2**
transversal   **121**, 122
trapeziums   31, **43**, 49
travel (distance–time) graph   **273**
trials   **216**
triangles   39–41
  angles in   39–40, 117
  drawing   131–3
  equilateral   23, **40**, 49, 135
  isosceles   **39**–40, 41, 42, 49
  properties of   49
  right-angled   **40**, 41
  scalene   **40**, 49
triangular prisms   140

U

union of sets   **279–80**
units, metric   99–100, 101, 161
unlikely   208, 209

V

variables   20, 111, 175, 221
Venn diagrams   279–80
vertex   36, **43**, 122, 139,
                              140, 192
vertical lines   225
vertically opposite
  angles   **36–7**
vertices   36, **43**, 122, 139,
                              140, 192
volume   **167–9**
  cubes   169
  cuboids   168–9

W

websites   73
whole numbers   6, 71

X

x-axis   **193**, 225
x-coordinate   194, 225

Y

y-axis   **193**, 225
y-coordinate   194, 225
years   266

Z

zeros   75, 76, 78